French Children Don't Get Fat

Marisa de Belloy

Copyright 2006 Marisa de Belloy
All rights reserved.
ISBN 978-0-6151-3765-0

For Maxime and Madeleine

Table of Contents

Prologue i

American eating habits…and those of the French 1

Taste Basics: Nature vs. Nurture 9

Why Taste isn't as simple as it seems 17

The Origins of Taste Classes 23

Learning to Taste: Class Basics 25

Beyond the Classroom: Teaching Taste at Home 49

All the Ingredients for Success 69

More fun and games 71

"Flavor is similar to love, something which many of us talk about rather readily, but which few of us care to define."

W. Gorman

Prologue

Eight months pregnant and loaded down with groceries from one of my favorite Parisian supermarkets, I plopped down on the bus planning the meal I was going to make that evening. At the next stop, a group of rough-looking teenage boys boarded the bus, filling up the seats next to and around me. I wasn't in a rough area, but had heard stories about women being harassed on public transportation and instinctively pulled my packages closer and looked out the window to avoid eye contact.

The boys were bantering in vulgar slang about a party they had recently been to when I suddenly heard one inquire, "Hey, do you smell that? It smells like basil!" The others chorused, "Yeah, it does." Basil plant sticking out of my supermarket bag, I turned back to reassess these kids. Nope, they were still the same rough-looking teenagers from before, but unlike their American equivalents, these kids knew their herbs!

I related the story to my French husband that night. Not at all surprised, he proceeded to tell me about the taste lessons that French children undergo, both at home and in specialized courses. French children are taught to appreciate food and the eating experience and to search out refined and interesting tastes. They learn about different foods, different tastes, smells and textures as a way of enhancing the pleasure they take in eating. The result is they think of a *real* treat as an exotic type of honey to be slathered on fresh, crusty bread or a very well-seasoned piece of meat, not a factory-made devil dog or a bigger-is-better fast-food monster burger.

All I can remember about my taste education, despite parents who enjoyed good food and a mother who liked to cook from gourmet cookbooks, is the four food groups with their impossible-for-a-child-to-care-about accounting of servings. The rest I learned from practice: For example, I learned that if I badgered my mother enough for Burger King French fries, I'd get them every so often.

As a kid and still today, I love cake, cookies and giant plates of ravioli, the more the better. I've got American eating preferences in a land where people both enjoy their food more and are thinner. I find I can read all the "Understanding the French" books I like - yet I still crave gummy bears. Now that I have my own child, however, I want to make sure that he doesn't inherit my questionable eating habits, that he doesn't have to struggle and stress about calories and extra weight. So, I set out to understand how the French really do it – how they raise children who, well, eat like French people! This book is the result of what I found.

American eating habits...
and those of the French

"Pour bien manger, il faut manger avec plaisir."
"To eat well, you need to take pleasure in eating."

French saying

    The first week after moving to Paris with my new French husband, I attended an "Introduction to France" for American women. Among the speeches and workshops on French healthcare, bureaucracy and where to find what, a nutritionist spoke about healthy eating in France. Her first order of business was to list common French foods and give the number of calories each one had. She started with *pain au chocolat*, a common French treat made from flaky pastry and chocolate. She read off the name, glanced down at her official-looking clipboard and, after a dramatic pause, proclaimed "550 calories!" The crowd of American women gasped in horror, many no doubt thinking of the tasty *pain au chocolat* they had picked up at the local bakery on the way over.
    Would a group of French women react this way? Never! In fact, they'd never even agree to attend a lecture like this in the first place! Why? The French don't care about calories! They eat what tastes good, they become satisfied, mind and body, and then they don't eat any more until they get hungry again.
    Not so for me. I've always been obsessed with weight. I started dieting at twelve, but was aware of calories and the guilt associated with eating too many of them well before that. My friends and I would alternate, starving ourselves and pigging out on junk food, whole bags of Oreo Double Stufs at a time, sometimes with canned frosting added on top. I was torn between the food philosophy of my old-school Jewish grandmother whose two favorite mottos were "eat, eat, eat" and *"es, es, es,"* ("eat, eat, eat" in Yiddish) and my aerobicized, calorie-obsessed mother. Food was something to be stuffed inside you in large quantities or avoided for as long as possible with all the will-power you could

muster. The idea of enjoying complex flavors and eating foods that tasted good beyond the first sugar rush simply never entered the picture.

Despite different backgrounds, Americans often have a similar story to mine. We quickly learn to eat double-sized take-out meals, on-the-go and at any time of day. We often eat with the television on or alone since the family can't find the time to eat together. We use food mainly to "fuel up" - and talk constantly about dieting. We stress and struggle and binge and purge. Our children take all this in and recreate it when they become adults, helped of course by the fact that junk food is available everywhere, at all times of day, and that very few children, male or female, learn to cook. Children across the country grow up having an abnormal, stressful relationship with food and many – 16% to be exact, up from 5% in the 80s – wind up becoming obese. Obesity starts early (some doctors say the critical age is three to four years old) and once you're obese, it's very hard to go back to being slim. How can we stop this downward spiral? How can we make sure our children start out right with a healthy relationship to food?

By now most people know that the French eat small portions and don't pig out on candy bars or extra-large frozen yogurts with sprinkles. However, before you or your child can eat "like the French," you need to understand the reason why. Here's the secret: The French love <u>good</u> food, not fat-free, low-cal, microwavable, chemical-laden frankenfood, but the honest, old-fashioned butter-, cream- and calorie-laden stuff. What's more, they don't feel guilty about eating. In France, it's considered right and natural to like food and enjoy eating it. Instead of the impromptu competition to eat as little as possible that often takes place among a group of American women eating lunch together, French women eat what they think will taste best. French people have a well-developed sense of taste and enjoy all aspects of a nice meal without worrying about calories, nutrition, or health scares. In a survey conducted by Paul Rozin, a famous American psychology professor who researches food, 75% of French women agreed with the statement "Enjoying food is one of the most important pleasures in my life," whereas only 42% of American women did. In short, in France, meals are wholly pleasurable events in which people spend time enjoying the food and the company.

To eat healthily, you need to have a well-developed sense of taste and an appreciation for food. When you're really able to appreciate taste, you're naturally drawn to the more complex tastes found in healthier foods. It's also a virtuous circle: the more you train your taste buds to like new, complex foods, the more you're drawn to a wider range of healthier foods. The best part is that when you can really appreciate your food, you don't need a super-sized portion to be satisfied.

Like other skills such as sports and languages, taste is a skill best developed at an early age. New research shows that children develop fixed food preferences and eating habits between the ages of two to three. Older children are also generally still flexible, but as any adult who's tried to embark on a new diet recently can tell you, once you're grown up, changing your eating habits is next to impossible.

So, where do we go wrong – and where do those paradoxical French get it right?

Stress!

For the French, eating is a sensual pleasure that they enjoy wholeheartedly. For us, it's a major source of stress, guilt and anxiety. Thirty percent of American college women say they would be willing to take a nutrient pill rather than eat so they could avoid the stress! A third of us alternate between binge eating and draconian diets. The skyrocketing rates of eating disorders among younger and younger children of both sexes show that we pass these habits on to our children. Some experts blame a variety of health problems on losing touch with the pleasure of eating. People who are comfortable with eating are able to relax and enjoy a wide range of foods, while those who are not lose control over their food and wind up suffering from stress.

We Americans rely increasingly on experts, doctors, magazines and advertising to tell us what to eat. Food has become so depersonalized that we no longer know how to taste and eating has become a scary and stressful minefield of "should-dos". If it's not calories that hold our attention during meals, or the workout we're going to do afterwards to get rid of them, it's the latest health scare or craze making us focus on not eating too many French fries or eating a lot more kelp. Help! It's no surprise we increasingly prefer to stick with a few comfort foods. While the French begin each meal wishing each other "Bon appetit" (I wish you a good appetite), Americans may as well start with the opposite: "I wish for you a small appetite so you will not ingest too many calories nor anything that could be potentially dangerous, cancer-causing or contrary to your personal beliefs."

Junk food, and lots of it

To make matters worse, much of the food we eat is mediocre. It's often industrially-made with artificial ingredients and lacks anything fresh. Our focus is on convenience and speed. Those years of Twinkies and fried chicken on-the-run have dulled our taste buds to the point where we crave only the immediate pleasure of simple sugar and carbohydrates. Many of us never really learned to taste our food in the first place, starting right in at an early age with pizza and French fries several times a week. I recently saw an article in the *Wisconsin State Journal* in which the author ponders whether blackberries have lost their taste over the years, or whether we've simply lost our ability to taste them. Both are probably true. We demand convenience and choice in our food to such an extreme, that we're served blueberries in the winter and "gourmet" porcini-mushroom linguine out of the freezer. Of course they don't taste like the fresh, ripe berries of summer or homemade pasta sauce! And, we're used to it! Many of us can no longer even tell the difference between good-tasting food and average-tasting food, since we eat so much of the average-tasting stuff.

Unlike Americans, the French eat a very wide variety of foods. Look at any true French cookbook and you'll find dozens of different kinds of cabbage, beans and mushrooms. As my husband never tires of telling me, French cuisine is one of the two oldest and most varied cuisines in the world (along with Chinese cuisine). Comparing my French cookbooks to my American ones reveals that the French are prepared to eat a much wider range of things than most Americans, like snails, pigeons and sweetbreads, and, surprise, surprise, many are vegetables and other complicated tastes that one needs to learn to love. The French also take a lot of care in their cooking to bring out subtle flavors by adding just the right spices or seasonings, and they delight in being able to appreciate them. I'll never forget my husband coming home all excited because in a local market he'd found his favorite type of salt. (Yes, salt. I challenged him to a blind taste test that night and, believe it or not, he was able to rank three different types of salt by their quality.)

The French are not adventurous eaters in the American sense. They mainly like French food, but, fortunately for them, French cuisine is vast and varied. Sushi is still considered terribly exotic, much more so than in the US. However, what I've seen among most Americans who consider themselves "adventurous" eaters is that they seek out the American equivalents in whatever cuisines they're eating: in Thai restaurants, pad thai (thick noodles in a sauce laden with peanuts) is close to spaghetti, in Indian eateries, chicken tikka masala (chicken in butter sauce served with heaping portions of rice or bread) is similar to all-American chicken and rice, and in Japanese establishments, tempura (batter-dipped vegetables or shrimp), is that good old American favorite, fried food. Children are even less adventurous: I still remember my sister and I on a family trip to Germany ordering *weiner schnitzel* (fried veal, German style) or spaghetti at every meal to avoid having to try anything more exotic.

The truth is Americans eat the same foods over and over again. Most cultures eat a wider range of foods than we do and adventurous American eaters are becoming few and far between. The number of Americans who rated themselves 'very likely' to try new foods has dropped from a narrow 27% in 1987 to just 14% in 1995. A survey by anthropologist Jane Kauer found that nearly half of Americans eat the same thing for breakfast every day – how boring! The foods we eat are also the wrong ones. Many Americans' whole diets are made up of "fake" foods, foods created entirely from artificial ingredients.

Sugar, sugar, sugar

To compensate for this lack of taste in our pre-made and genetically modified foods, we just add sugar. Labs all over America are working to develop new and cheaper ways to sweeten everything we put in our mouths, from our foods to our mouthwash. We're so overpowered by sugar that we can barely taste anything anymore, and instead of backing away, we continue to invent more and more ways to make food sweeter than anything found in nature. The problem is that the taste of sugar overpowers the taste of other foods, so the only thing that tastes good is the sugar! So, we eat more and more sugar to satisfy ourselves, which keeps us hungry (since six cupcakes are not going to satisfy you that way a complete meal will!), overweight and unhealthy.

Knowing how to eat and loving food

Above and beyond just the food that they eat, the French know how to eat. Enjoying the whole food experience is where the French really excel – and we Americans come in last. Dinner in France, even when it's just family, is a grand social affair. Tablecloths, cloth napkins (gasp!), nice plates, a calm setting and people getting down to the business of interacting and enjoying their food. This and only this is considered eating in France and no one would give it up for a burrito on-the-go.

We Americans are just the opposite. Today's average American woman comes into contact with food 20 times per day, but she eats alone. Not only is eating by yourself not very much fun, but it also encourages binge eating and a focus on a few unhealthy comfort foods. Few people have the inclination to prepare even a simple meal for themselves when they're alone, so they often just order in, or "pick something up."

Even when we Americans cook at home, we don't tend to put much emphasis on making the whole eating experience a pleasant one, with varied, interesting food and a calm, family setting. Time constraints, stress and outside activities jostle us from beginning to end and make us inclined either to zone out during dinner – perhaps by putting on the TV, checking our email or reading the mail – or to rush through it to get to the "real" relaxing part of the evening. Many of us think of enjoying a meal as something to be done occasionally in a nice restaurant.

In contrast, French people have a love affair with food. Although they only eat at mealtimes, the French talk about food incessantly – where it is grown, which region is the best for which type of food, how it is farmed, how it is prepared, the best way they ever ate a certain dish, mistakes made in cooking, etc. The majority of French conversations, particularly over any meal, are about food. As a foreigner, you quickly develop a food vocabulary more vast than the one you had in your native language. Great chefs and foodies even attain celebrity status here: The 2002 death of baker Lionel Poilane, son of the inventor of a particularly good type of bread, received national news coverage!

Talking about good food raises awareness and enhances the desire to eat and try new things. It gets you thinking about what you're eating and interested in experiencing new, more complicated tastes. It also promotes a healthy attitude towards eating. (That doesn't change the fact that you'll need to have decent food to begin with, though. While we can discuss a good home-cooked meal for awhile: how it was made, where the products were bought, who prefers what dish, etc., there's not much to say over a fast-food hamburger.)

Talking about food is so integrated into French culture that many of the most common French expressions have to do with food. So, even when the French are not talking about food, they are still talking about it! Some of my favorites include:

AMERICAN EATING HABITS ... AND THOSE OF THE FRENCH

Expression	Literal translation	Figurative translation
Mon chou	My cabbage	My sweetheart
Mon poulet / Ma poule	My chicken (male or female)	My darling
Pédaler dans la choucroute	To pedal in the sauerkraut	To try but get nowhere (like "running on ice")
Prendre quelque chose en pleine poire	To take something straight in the pear	To get hit smack in the face
Tomber dans les pommes	Fall into the apples	Pass out
Appuyer sur le champignon	Push on the mushroom	Accelerate
Raconter des salades	To tell salads	To tell tall tales
Se prendre une châtaigne	To get hit by a chestnut	To get punched
Avoir un pépin	To have a fruit seed	To have a problem
Ne pas être dans son assiette	To not be in your plate	To seem unwell or uncomfortable
Se prendre une saucée	To get covered in sauce	To get drenched
L'appétit vient en mangeant	Appetite comes with eating	The more you have of something, the more you want
Être haut comme trois pommes	To be as tall as three apples	To be short
Être comme un coq en pâte	To be like a rooster in dough	To be well taken care of / To be comfortable
Être soupe au lait	To be milk soup	To have a quick temper
Mettre les petits plats dans les grands	To put the small plates on the big ones	To go all out
Tomber comme un cheveu sur la soupe	To fall like a hair in the soup	To show up at an inconvenient time
Avoir du pain sur la planche	To have bread on the chopping board	To have a lot of work to do
Mettre de l'eau dans son vin	To put water in someone's wine	To stop someone from being too extreme
Avoir la pêche	To have the peach	To feel energized

Relax!

When you're relaxed and happy about your food, your children will be also. Take vegetables. When it comes to veggies, American parents often shoot themselves in the foot in a way that French traditionally parents do not. Focused on calories, Americans, often unconsciously, pass on their food neuroses to children early on. My mother was always on one diet or another when I was a child and I remember her often eating just a plate of asparagus for dinner, "because it's a diuretic." What did I learn? That one eats asparagus because it stops you from retaining water and makes you momentarily thinner. Great for three hours before the prom, but not for a lifetime of healthy eating habits. What did my husband learn, growing up an ocean away? That one eats asparagus because, prepared with a little melted butter, it can be delicious.

Ok, so the French have a healthier attitude towards food than us Americans. Unless you've been living on the moon, you've probably heard this before. But, here's what you may not know:

1. Eating habits and food preferences are learnable, particularly for children, and from as early as two years old.
2. Regardless of whom you are or what country you live in, you can teach your children these healthy French habits.

… and all this with relatively little change or stress!

What few American parents realize is that children's palates need to be developed much as we develop their minds with education or their bodies by teaching them a new sport. The ability to appreciate complex tastes comes with practice. The same goes for learning to enjoy meals at mealtimes and not stuffing yourself on Cheetos at three o'clock in the afternoon.

Doing this means making a few simple changes to your family meal routine and making eating into what it should be: a fun, interesting experience for your children. It also means avoiding a few of the standard traps that we parents unwittingly fall into.

Whoever you are, it's not too late! By teaching your child to taste and to appreciate his or her food, you too can raise a child who eats *à la française*, the French way, in a healthy and unstressed way. Why would you want to? Well, first of all, imagine your child growing into an adult who eats a healthy, varied diet by choice, who takes pleasure in good meals and who never has to worry about calories or nutrition. This is a great gift to give a child – one less source of stress and one more source of pleasure in a world which often has too many of the former and not enough of the latter. Remember, French children (and American children who follow this method) don't get fat!

Taste Basics: Nature vs. Nurture

"Pour connaître les gens, il suffit souvent de regarder ce qu'il y a dans leur assiette."
"To really know someone, you often only have to look at what he has in his plate."

French saying

Before getting into how the French do it, let's have a quick look at what we've got to work with. Humans are omnivores, ready to eat most things, which is a good start. We're born with few natural likes and dislikes for foods. Our culture and environment teach us what is good and what isn't and help us to create food preferences. From here, we can start thinking about developing our palates and moving into eating a wide range of healthy foods.

In the beginning....
The main concern of a newborn child is to satisfy its hunger. Normally, this is a happy (if sometimes ill-timed!) process for both mother and child.

Baby humans are programmed, however, with at least one basic rule of survival. Sweet foods, the sweetness signifying high caloric value, are good, while bitter foods, often signifying poisonous foods, are bad. Not surprisingly, mother's milk has a sweet flavor.

These instincts and the way we react to them are universal. Studies conducted by a famous French child psychologist show that when young babies are given something bitter tasting, they make "the bitter face," the same face you would naturally make if you bit into a bitter grapefruit or drank acidic coffee, whereas when they taste something sweet, their face relaxes with hints of a smile. By 18 months, babies are already capable of making this bitter face in any non-food situation that displeases them, reinforcing the "bitter = bad", "sweet = good" reflex. This desire to eat sweet, immediately satisfying foods stays with children until they are taught, through education and environment, to expand their food repertoires and develop their sense of taste.

Although we're naturally programmed to eat a wide variety of foods, we're also programmed to be wary of new foods in order to avoid poisoning. This explains why children start out liking only foods that they know and slowly they expand by association to foods that are similar and so on and so on. Adults continue to do so in a less obvious way. Finding that frogs' legs, for example, "taste like chicken," is a reassuring way to classify an unknown food into something that is both new and familiar at the same time. For you this means: The more foods your child has tasted, the more comfortable he'll be with a varied diet.

<u>Taste hardware and supertasters</u>
When it comes to taste, we are all born with similar "hardware," but it is arranged in strikingly different ways. Let's start with some taste basics. Starting from the third trimester of pregnancy, we all develop taste buds, which send messages to our brains with information about one or several of the five tastes. The five tastes include sweet, salty, bitter, sour and the more recently discovered *umami*, which was named by a Tokyo chemistry professor for "deliciousness" in Japanese. Its taste is similar to chicken soup.

These taste buds are stimulated in different combinations, creating what we call flavor. In fact, it's extremely rare to experience one taste in isolation. Instead, flavor occurs at the intersection of these five tastes. Even something as seemingly mono-taste as Nutrasweet, for example, is both sweet and a little bitter.

Combined with taste is the all-important sense of smell, activated through both the nose and mouth, and together they make up the flavor of a food. Believe it or not, smell makes up 90% of flavor compared to the 10% made up by taste. Try holding your nose and biting into a lemon. What you'll experience is only the taste of sour, but not the smell that transforms that sour taste into the flavor "lemon." The same goes for when you have a cold: It's much harder to get the flavor of the food you're eating. There are even flavorings like vanilla, for example, which have no "taste" in technical terms, only a smell, but since smell and taste are so closely related, we don't even notice.

The similarities stop there. Just as no two snowflakes are alike, no two people have the same configuration of taste buds. For example, twenty-five percent of us are "supertasters," born with more taste buds than others. Supertasters have a better, more intense perception of flavors. They can experience up to twice the impact of straight sugar, for example. They may also wind up being pickier eaters who are more likely to avoid bitter-tasting vegetables since they're more sensitive to taste. If your child is a particularly difficult eater, she could be a supertaster. Interestingly, East Asians and women are more likely to be supertasters than other people, although modern science does not seem to know why. (One hypothesis is that women are programmed to protect their unborn children and need to be sure to taste possibly harmful foods right away. For the East Asians, it's anybody's guess.)

Who are you?	% of Population	How to Tell
Supertaster	25%	You love chocolate and hate raw broccoli because it's too bitter
Normal taster	50%	You are happy to eat a wide variety of foods
Non-taster	25%	You don't find a lot of difference in how foods taste and aren't that interested in eating

We're also all born with different sensitivities to smells. This plus our different taste-bud configurations gives us different thresholds for tasting foods. Scientists who study taste have defined three thresholds:
- The ability to taste something (The absolute threshold)
- The ability to recognize what you're tasting (The recognition threshold)
- The ability to decide whether or not you like what you are eating (The preference threshold)

These thresholds influence how we like our food. A small amount of salt may seem way too salty to one person while to another it hardly comes through at all. Some people find that a half packet of sugar in their coffee gives it enough sweet flavor, while others need two or three packets to get the same sensation. Children have much lower thresholds than adults for deciding whether or not they like what they are eating. This could explain why they're quicker to say they like a certain food and quicker to reject other foods. The good news is these thresholds can change with age or training. This means that trying to appreciate the same food with less sugar, for example, a fundamental principle of many diets, can work if you stick to it. It also means that your children can also learn to prefer eating less sugar!

Your taste, my taste

Different foods actually taste different to different people. If your child doesn't like a food that you like, this could be because it tastes completely different to him than to you. Although everyone agrees that sugar tastes sweet, and salt salty, no one agrees on flavor. That's why you may love mushrooms on your pizza and your best friend hates them.

The same goes for smells. Adults find they can love the smell of, say, sausages or burning rubber, while their friends or family find these smells nauseating. If someone lent you his eyes, you would see pretty much the same things you're seeing now. But if you borrowed someone's nose and mouth, you'd be shocked at how different foods tasted.

But don't despair just because your kids taste things differently. Biology is on your side. The majority of us are programmed to eat a mostly balanced, satisfying diet. Studies conducted with orphans and children in hospitals showed that they naturally chose food that kept them well-fed. Even babies know how to regulate their intake of food: when mother's milk is diluted, they drink more to get what they need. And, anyone who's been through a few days of eating only junk food or traveled in a country where uncooked foods are not recommended for contamination reasons will recognize that craving for "something healthy."

How we start to develop our likes and dislikes

It starts even before birth! Already in the womb, children are exposed to the likes and dislikes of their mothers. Babies whose mothers drink carrot juice during pregnancy are more likely to enjoy eating carrots later on. Experiments with animals have shown that we all experience the same phenomenon. Even baby rats whose mothers ate more garlic were more prone to like garlic.

Breast-fed babies whose mothers eat a wide variety of foods are more likely to be open to trying new foods than formula-fed babies, probably because babies experience the residue of flavors their moms eat in breast milk. Studies show that breast-feeding babies whose mothers eat garlic will suck longer, perhaps to continue exploring this new flavor.

After that, though, parents' direct influence on their children wanes. Most children's tastes do not greatly resemble their parents', meaning that children really do make their own choices where food is concerned. Your job is to teach them to develop their taste so they make good choices, even if they don't correspond exactly to the foods you like.

TASTE BASICS: NATURE VS. NURTURE

<u>The role of culture</u>
What we do know is that humans can learn to like any food. Imagine two children with the same exact taste buds, one growing up in the US and the other in China. Each would have a completely different set of favorite foods. Think about it: Almost anything that has nutritional value is consumed by some culture somewhere in the world, including plants, insects, and elaborate confections of things we Americans would be hard pressed to eat. *The Times* of London ran an article a few years ago about a new bestselling product offered to sweltering Tokyoites: Raw horse flesh ice cream. Japanese children may have loved it, but just try getting your child to eat it! Culture also accounts for many of the things we don't like. It teaches us that foods not traditionally eaten, such as, for Americans, raw lamb, snails, cockroaches, etc., are disgusting.

These culture-based preferences are strong. Informal evidence indicates that food preferences are the last thing to change among immigrants to new countries. The plethora of ethnic restaurants and food stores in any immigrant community shows that one of the first reflexes of immigrants is to recreate a food environment similar to the one they left. I've also found this to be true. Living in, Paris, the food capital of the world, I still regularly make pilgrimages to The Real McCoy, an American store in the 7th *arrondisement* (section of Paris) for peanut butter or Oreos. The American community in France is rife with information on how to recreate American flour (by mixing two types of French flour), where to find a decent American-style Sunday brunch and how to prepare a Thanksgiving dinner in a country without a love of turkey and cranberry sauce.

Many cultures simply have no interest in trying other foods. My French husband tells of hosting a group of Chinese business associates for a month who quickly turned up their noses at the number of nice French restaurants he had booked for them during their stay. Instead, he wound up driving them to Chinatown, where they found the foods they wanted and cooked them themselves.

<u>Eating by Age</u>
Why is it so important to make sure your child eats well? As they grow up, children establish the food preferences and eating habits that they will keep for the rest of their lives. Like anything else, it's much easier to put your child on the right track from the start than for her to try to radically

change her eating habits at thirty. A child who is obese at 15 has a 75% chance of remaining obese as an adult. On the other hand, a child who has developed good eating habits and an appreciation for food is much more likely to grow into an adult who continues on this course.

Regardless of what you do, most children develop their eating habits in very similar ways, so let's take a quick look at these cycles:

- Before two years old: Most children will eat most foods if they're hungry.
- After two: Kids become picky: Only one-fourth of children will eat something new the first time it's presented.
- By 5-6 years old: Children start to develop real likes and dislikes
- 10 years old: Children become even more finicky and start to reject foods they were happy to eat before
- Late teens: Children begin to eat like adults, having established the food preferences and routines that will see them into old age

The key to understanding how children eat before the age of five or six is this: Children like what they recognize. "I like it" means "I know it" and "I don't like it" means "I don't know it." Children judge foods by the degree of familiarity and the degree of sugar (a new sugary candy will always be accepted!)

What you need is to make foods seem familiar. The only problem for you as a parent is that familiarity is hard won! A small change in appearance, or sometimes even a change of table setting will render a dish "unfamiliar" and therefore inedible, a phenomenon which can be frustrating and baffling for parents. I once read the internet posting of a desperate mother whose son would no longer eat mashed potatoes because she had once dared to hide a pea in them. Although any parent will recognize these tendencies, a study with rats proved it: The study showed that while adult rats were willing to try many new foods in small quantities, baby rats would rather starve than eat something new!

Before about six years old, children don't think of food in the same way as adults. For example, they have no sense of what adults consider disgusting in food. Most children younger than six are willing to drink a glass of apple juice with a comb in it, even if the comb looks like it's recently been used!

Taste Basics: Nature vs. Nurture

Even if you've finally gotten your five or six year old to accept a number of familiar foods, you're not out of the woods yet. At 10, a new rejection phase begins and once adolescence starts, a whole new set of tastes develops, partially influenced by peers, a desire to stand out among family members and fit in among friends. It becomes particularly hard to get your child to eat reasonably healthy foods as he is bombarded by unhealthy options which seem "cooler."

Girls can also be more difficult than boys. They're more likely to know what they want and reject things that don't seem appealing. They are also more sensitive to sweets and more likely to seek them out. During puberty, girls undergo a hormonal change which makes them even more dessert-crazy.

Age also plays a more general role in taste preferences. Children are more naturally attracted to sweet and starchy foods that satisfy quickly since they're not good at delayed pleasure. Babies can't get enough of sweet foods, but they, and slightly older children, also often like strong sour tastes, the kinds that would make an adult cringe. A study found that children ages 5 to 9 actually enjoy the taste of concentrated citric acid. I once found my 2-year-old son happily sucking on a lemon wedge that I'd left on the counter, as if it were a piece of candy. What I've since heard from other mothers is that this is not uncommon among small children!

Bitter foods, on the other hand, become more appealing to adults whose palates are more developed. (Another theory that could explain this is that older people are naturally attracted to foods which contain cancer-fighting agents, foods which also happen to be bitter like broccoli rabe and kale.) While the human mouth has only a few different types of taste receptors for tasting sweet, it has a wide variety that react to bitter, and so bitter may simply be a more complex taste that it takes years to learn to enjoy.

With such a complex set of biological, environmental and cultural factors, you have to wonder, "Can a person's taste be influenced?" The answer is, "Absolutely." Almost everyone has learned to like a food that seemed unappealing at the start: sushi, coffee, spinach, etc. This is how cultures continue: baby Mexicans learn to eat chile peppers, baby French learn to eat oysters and, conversely, baby Indians are taught that eating steak is repulsive. All over the world people learn to like things that don't seem initially appealing. In fact, it's generally only in rich countries, among people who have the financial means to buy different foods that we find so many so-called finicky eaters!

It's up to <u>you</u>. You can educate children in taste just as you teach them to play softball or play piano, knowing it's a pleasure they will be able to appreciate for the rest of their lives. In doing so, you can also teach them to accept all sorts of healthy, tasty foods. Studies show that children accept new foods which are presented to them repeatedly (5 to 10 times) and that is what makes them into adults who eat a wide range of foods (even if they go through a period of eating only Cheerios and peanut butter!). After that, it's a virtuous circle: The more open we are to trying other foods, the more likely we are to continue trying new things, increasing the probability of finding new foods to like.

Eating habits are also passed on from parents to children, even if specific likes and dislikes are not. Trying different things, eating only at meal times, enjoying foods with complex tastes like vegetables are all things children learn from observing their parents – and they learn them early (starting at 2-3 years old). Just because your child can't yet talk in full sentences doesn't mean he is not taking in everything you do and creating his own personality and habits around it!

WHY TASTE ISN'T AS SIMPLE AS IT SEEMS

We make assumptions about what we eat, even before we eat it. Most of the time we've already decided whether the food will taste good before even putting it in our mouths. In my experience with visitors here in Paris, Americans find that everything they eat here is excellent, because they expect it to be excellent. I've seen this phenomenon even with the most basic sandwich or hamburger – foods that we Americans excel at making and the French, well, just don't. Researchers have shown that even wine experts, whose job it is to recognize good wines, can be fooled by being served excellent wine in a cheap bottle or vice-versa.

In the assumption game, children are often the worst offenders. More so than adults, they judge most foods by the very first look they get at them. What's more - the moment they get a hint from you that something is good for them (read: tastes bad) or that you yourself don't actually like it, they'll find it disgusting regardless of how it actually tastes. Your job, as we'll see later, is to make sure the assumptions they make about foods you want them to eat is that they will be good.

Taste vs. Flavor

But, back to taste manipulations. "Taste" is just what happens when food hits our tongue. The rest – all the other sensory information that helps us to enjoy or dislike our food – combines in what researchers call "flavor," and these other sensory elements play a much larger role than we realize in determining what we like and dislike. American doctors advise their patients with taste disorders, for example, to maximize food texture, aroma, temperature and even color, and the patients then report being able to enjoy their food. What does this have to do with you? Well, even if your children don't like the taste of broccoli, all hope is not lost! Think of getting them to like the flavor – how it's served, what it looks like, smells like, feels like, what kind of setting they're eating it in.

Truly being able to appreciate your food means appreciating flavor, not just taste or one single sensory element of the food. It means enjoying everything about that food, including how and when you eat it. When we aren't able to appreciate flavor, we get sucked into food marketing campaigns and find it harder to recognize what's really good and what we're just being tricked into thinking is good. Let's look at a few examples:

<u>Manipulating color</u>

Although the difference between taste and flavor may be new to you, it's not new to the junk food companies who market to your children. (Have you ever wondered why those candies stacked at a child's eye level in supermarkets have such bright colors?) Take soft drinks, for example. When researchers ask people to drink the same soft drink distributed into three different cups with only a change in that drink's color, the tasters routinely report that the sweetest soft drink is the one with the most intense color. You can use this to your advantage, though. Preparing foods in bright colors (even if you sometimes cheat by slipping in a little food coloring,) will make them instantly more attractive to your child.

The opposite is also true: Without seeing the color, people cannot always identify the flavor of a soft drink. Another study showed that, in a blind tasting experiment, only 30% of the people tasting a cherry drink correctly guessed it was cherry. The majority (40%) thought it was lemon lime! Even wine experts can find themselves unable to identify whether a wine is red or white if the color has been altered. So, here you are thinking your favorite soda is cherry or your favorite wine is a golden chardonnay and in fact, without seeing the color when you drink, you could drink something completely different and still get the same sensation. What does that mean? It means the drink you're drinking is not really that appealing in terms of taste, smell and general satisfaction. You're just being tricked by the color.

Color can also be naturally appealing or unappealing. Given a choice among colored jelly beans, each with the same flavor and concentration of sugar, most people choose the red ones over yellow or green. Red and then purple are judged to be the most sweet, probably because in nature shades of red and purple are associated with ripe and green, not ripe. Blue, on the other hand, a color not often found in nature, is often rated as repulsive, regardless of the taste that follows it. You've probably seen this phenomenon in your children as well. For example, you might find that while your children don't like green or yellow peppers, they may accept red ones. (Why not try a mix of all of them?) Understanding color can help you get your children eating a more varied set of foods.

Why Taste isn't as Simple as it Seems

The French understand the role of color. Finding an appealing mix of colors is considered essential to a good meal. I remember my mother-in-law being thrilled when a friend in Normandy brought over a variety of purple potato that she could use to make the evening's normal mashed potatoes look more attractive and interesting. Even the color on the plates should enhance the overall meal.

Believe it or not, the color of the room in which you're eating has also been shown to affect taste. People prefer sugary foods in red rooms, salty ones in yellow rooms, sour ones in green rooms and bitter ones in blue rooms. Ever noticed that many candy stores have red logos and decorations?

Manipulating sound

We also have certain expectations about how things should sound when we eat them. For example, volunteers who heard less of a crunch when eating a certain brand of potato chips believed the chips were soggy, despite the fact that they were exactly the same fresh chips they'd eaten previously. Once again, don't think the chip manufacturers don't know this!

All is not lost, though, for us consumers. Once we realize these effects on what we think of as our preferences in food, we can use them to our advantage, getting our children to like foods whose taste they initially reject simply by turning up the effects of other sensory information.

Temperature

Take temperature. As any great chef will tell you, temperature is a very important factor in how things taste. Have you ever noticed that apple pie is less sweet when it's cold? So, it's the day after Thanksgiving and you've got a few slices of apple pie left for dessert. Simply by warming up that pie, you can make eating it a much more satisfying experience, so that one slice will feel sufficient. Eat it cold and you may find you can go through three or four pieces without getting to that feeling of satisfaction. The same goes for your children.

I have a running debate with my husband over temperature. We order in a pizza, for example, and it arrives lukewarm. My instinct is to pounce on it as soon as I see it and his is to stick it in the oven for 10 minutes to get it back to the gooey, melting-cheese experience that is pizza at its best. Sometimes I win, we don't heat it up, I dig in and eat about two to three slices more than I should. When he wins, we savor the pizza and I go away feeling fulfilled after a reasonable two slices.

In fact, each taste has an "optimal" temperature, a temperature at which it tastes the best. They are:

Sweet: 99 degrees Fahrenheit
Salty: 64-72
Bitter: 50-59
Sour: Varies little with temperature

Adaptation

Then there's what they call adaptation. Our taste buds adapt to what we're eating and make it so we can't taste anymore. Maybe you've noticed that the first few bites of ice cream can be delicious and extremely satisfying, while by the 20th bite you barely taste a thing. Here's something that the French know instinctively: You're much more likely to be satisfied by a small portion of something good than an extra-large portion of the same food. Quite simply, half way through that jumbo-sized frozen yogurt, you aren't even tasting it anymore. The effect is only in your mind. The less complex the taste, the faster it's going to be "adapted." Sweet and salty foods, for example, go first. By the tenth chip from that bottomless chip basket, you may as well be eating cardboard!

Another French strategy against adaptation is to eat a variety of foods at any one time. If you have a few bites of several different foods and alternate among them, you can keep tasting and appreciating all of them longer, since every time you switch foods, your taste buds become "un-adapted." So, at the end of a meal you feel more satisfied than if you had eaten one food at a time.

Memory

A good memory of a favorite dish from your past may make eating that dish in the present less pleasant in comparison. Most people can remember a food their grandmother made for them and will gravitate toward that food without regard to other considerations. For me, it's chicken soup with matzo balls. Not knowing how to make it myself, I've tried it in a range of Jewish delis across the world only to be disappointed by the flavor's never quite being the same. Quite possibly, matzo ball soup is just not that good, but my brain still tells me it's one of the best foods on earth. (Even the French don't have a recipe to counter this one!)

<u>Language</u>
Language plays a major role in food enjoyment. Without the correct words to name what you are tasting, it is hard to appreciate fully the taste. Did we know we were tasting *umami* before the Japanese invented a word for it? Children are particularly limited in food language which limits their ability to appreciate food. It's no wonder the French talk about food constantly!

<u>Disgust</u>
At the other end of the spectrum, a food that once made you sick is bound to spark negative feelings that can ruin its natural taste for you. Disgust also plays a strong role with certain foods, including body-part foods like tripe, brains, liver and foods with strong flavors like olives and cheese, regardless of actual taste. Americans allow themselves to be disgusted by a wide range of foods; the French do not. Humans are also the only animals who are thrill seekers when it comes to eating: People who seek out the absolute hottest salsa, for example, are often looking for the thrill rather than the taste.

<u>Culture</u>
We also manipulate taste via culture. Take something as simple as the time of day. Already at four years old, children say they prefer a hot chocolate in the morning and pizza in the evening. A meal that would seem incredibly appealing at the end of the day can be off-putting in the morning - But this is purely a cultural creation! Some cultures, the Japanese for example, eat similar meals in the morning and evening, including fish and rice and other foods that a westerner would label as "dinner foods." The French have a very strict set of rules concerning what is eaten when – and this helps them not to overeat. In the US, we consider it our right to eat cold pizza before breakfast and breakfast cereal in the afternoon. Unfortunately, these relaxed rules get us eating whatever we want whenever we want it.

We all also have a learned understanding of what's eaten before what. Soup comes before the meal, salad either before or after, depending on whether you're American or French. Either one may taste "funny" if eaten out of the expected order. In the US and England, meat should be accompanied with two smaller side dishes, ideally one starch and one vegetable. In France, main dishes are often served with only one side dish. In most western countries, eating just a steak for dinner may feel wrong, even if Atkins-diet devotees would have you believe it's the way to a perfect figure.

And More…

The list of natural ways to manipulate flavor goes on. Sunlight increases taste and darkness decreases it. Perhaps this is one the reasons the French flock to outdoor cafes and terraces as soon as the first ray of sunlight comes out in Paris? Or maybe this is part of the reason they prefer to have their main meal during the day?

Background noise can also take away from flavor. A five-star meal eaten on an island in the middle of a noisy, busy highway would no longer be a five-star meal. Dinner with the television on is less tasty than dinner in a calm setting. (The French are fanatics about calm dining environments.)

Even more simply, foods taste better when you're hungry and worse when you're thirsty. The French don't eat between meals to make sure they are hungry by mealtime. If they're not hungry, they simply don't eat. They also drink a lot of water and eat a lot of high-water-content foods like soup and fruit at meals.

So how does all this matter to us parents trying to keep our children slim and healthy? We can manipulate the way food tastes in a way to make foods more attractive to our children. Comprised of subtle flavors and changing internal and external factors, food preferences are not as simple as a dislike of lima beans and a craving for Krispy Kreme donuts. Although I wouldn't necessarily recommend painting your kitchen blue to increase the likelihood of your children eating their vegetables (however, if it is blue already, more power to you!), by being sensitive to all the factors that affect taste: sight, sound, temperature, food order and surroundings, you can greatly influence what appeals to you and your family even without increasing your skills as a cook.

THE ORIGINS OF TASTE CLASSES

Only the French could invent *"les cours de goût"* or taste classes, practiced in primary and secondary schools across France. The credit goes to Jacques Puisais, the father of taste classes, who began his classes in 1984. Puisais's belief, now proven by science, is that taste can be learned. Starting from that premise, he created courses to help children learn to appreciate food by discovering and learning about new foods, including how to prepare them and how to talk about them.

If the French eat so well, why do they give their children taste classes? The French are not unaffected by the rise of fast food and eating on-the-go. A friend who works at Coca-Cola in France tells me that France is one market which always does well and lunchtime lines outside McDonald's, which continues to grow in France, can be commonplace. Taste classes are an attempt to formalize what the French do naturally, to make sure that fast food doesn't gain an even stronger foothold. The theory is simple: When we are able to taste, we naturally eat more healthily and avoid fast, fatty foods in favor of more complex tastes.

Puisais discovered the positive effects of taste education when he introduced a class of children to a number of different smells. What he found was that with a little bit of exposure, these children were more able to identify correctly certain smells much better than children who hadn't been exposed to the smells and better than the average adult. However, they didn't have the words to describe what they were smelling and resorted to phrases like, "That smells like a hospital," or "It reminds me of Grandma's house." Once he taught the children words to express what they were smelling, he had created a group of children much more sensitive to smells and able fully to appreciate them.

A few years later Puisais developed his first series of taste classes, including ten sessions of one and a half hours, dealing with everything from the five senses to tasting foods from the different regions of France. His goal was to make children aware of different taste sensations and enable them to critique their food and environment. He wanted to make children into intelligent omnivores, willing to try new foods and open to a wide variety of flavors, giving them a firm grounding in food as a personal and pleasurable choice. Puisais found that among children who had taken his class, many more were likely to say they were willing to try new foods. Over 95% of children said they would like to take the class again.

Most kindergartens and primary schools in France have some sort of taste education. Puisais runs an institute to offer classes to 7 to 11 year olds, but he has spawned a movement that is copied by most French schools across the country. Every year during France's "Taste Week," children aged 8 to 12 take classes given by famous and local chefs and almost all schools in France diversify their menus and run various activities that get children thinking about taste. The result is that most French children experience some kind of taste education in school.

Today, in France, ensuring the taste abilities of future generations is a major concern. For example, Jack Lang, France's former Culture Minister recently spoke about his view on taste classes and the role of the government. "Our goal is simple: Endow each student with a sound general culture in food which will be useful for him his life…" I find it typical of the difference between how we Americans eat and how the French eat that the US Government is more concerned with issuing guidelines telling us how many calories and servings we should eat of different types of foods, while the French government takes a serious interest in making sure its population enjoys its food!

Learning to Taste: Class Basics

"Le goût ne s'impose pas, il se transmet, par une série de petits gestes."
"Taste is not imposed, it is transmitted by a series of small gestures."

Jacques Puisais, founder of taste classes in France

Taste classes are a simple and fun way to keep children slim and eating healthily by:

- Getting them thinking about food and eating
- Giving them the vocabulary to talk about food
- Helping them to try a range of different foods

Why do the French bother to teach their children all of this? First of all, the more children understand about food, the more likely they will be to search out complex tastes, appreciate good food and reject substandard, "convenience" food – not because they're worried about calories, but because it simply doesn't taste as good as the real stuff. It's just like enrolling your child in a class about art. At the beginning, any cartoon painting of Mickey Mouse will look great to him – bright colors, a subject he recognizes – but after a few classes, he will begin to appreciate more advanced art for its subtle lines and messages. The same goes for sports. A true baseball aficionado sees the beauty in Roger Clemens' pitch, while the novice sees only which team wins.

The French start their children off trying a wide range of foods. Remember that familiarity with a food is key to getting your child to eat it. It's also key to appreciating food. No one can really appreciate the same few foods over and over again since part of the real pleasure in eating is trying new things. It also helps to find new favorite foods. Ok, so maybe your child will always hate tuna fish, but perhaps salmon will become his new favorite?

Let's have a look first at an outline for one of Jacques Puisais's courses in developing taste. It includes ten lessons of an hour and a half each on:

1. Using the five senses to experience and talk about food
2. Understanding the four (or five) tastes
3. Preparing a meal
4. Smell
5. Sight
6. Touch
7. "Taste under attack"
8. Regional cuisine
9. Recap
10. Top notch food and closing ceremony

At the end of each class description, you'll find a "How you can try it at home section" with suggestions on how to adapt these classes to your children. For the most part, these exercises are suitable for children from three and up. Where that's not the case, you'll find a separate adaptation for other ages.

Learning to Taste: Class Basics

Class One: The Five Senses

So, what does it really mean to like a food? When we say, "This tastes good," we might really be saying any combination of:

- "This smells good"
- "This looks good"
- "It has a pleasant texture"
- "It has a nice crunch"
- "I was really hungry"
- "The ambiance here is very nice"
- "I like the colors of the food and the surroundings"
- … and many, many more

When we understand these other impacts, we can help our children understand and enjoy food. But first, children need to have the vocabulary to talk about what they're experiencing.

The teacher places a few foods in front of the class and guides them to talk about what the "food is saying." For example, "I'm a green banana and I don't give off a smell. That means I am not ripe," or "I'm a fresh baguette - I crackle when you eat me." "I'm a red apple. I'll be sweeter than a green one."

All foods are fair game. "Who can name a food that makes a loud sound when you eat it?" "Do you prefer your soup hot or cold?" "What's the stickiest food you've ever eaten?"

Then comes the game. The teacher brings out a large box of *pâte de fruit*, a dense, sweet, sugary candy with the consistency of gum drops that comes in a variety of intense fruit flavors. Each child selects one and studies it: What does it feel like? How does it smell? Does the smell reflect the color? Is it what you were expecting? Is it strong or weak? They pop the candy into their mouths. What sound did it make? What does it feel like in your mouth? How does it taste?

Children begin to see what it is to understand a food. They also see how their preferences differ from others in the class. Maybe a child chose the green candy when the majority of the class chose red. This child starts to understand that he has a choice and distinctive preferences. "The candies are all good, but I prefer the green ones."

French Children Don't Get Fat

<u>How you can try it at home</u>
Whatever their ages, talk to your children about food. Get them thinking about how a food looks, smells, feels, etc. Ask them what kinds of food they prefer and how they prefer them: Hot? Cold? Mushy? Firm? Purple? Red? The following chart can help.

<u>Words to Describe the Foods We Eat</u>

<u>Sight</u>
Shape: Round, square, oval, spherical
State: Liquid, solid, gas
Consistency: Thick, heavy, light
Appearance: Clear, cloudy, shiny, dull, rough, smooth, etc.
Color: red, green, multicolored, etc., dark or light

<u>Sound</u>
Crackly, crunchy, loud, quiet, squishy
The sound of food being served on a plate or in a bowl
The sound of eating food

<u>Smell</u>
Fruit smells, flower smells, spice smells, the mix of smells that one can experience in one food
Sharp, odorless

<u>Taste</u>
Sweet, salty, sour, bitter
Tastes like....
Spicy, tart, ticklish, metallic, biting

<u>Touch</u>
Hard, soft, sticky, rough, greasy, oily, soft, smooth, velvety, stringy, creamy, etc.
Temperature: hot, cold, room temperature

---

1. Describe different foods as you serve them. "Here's a tomato. It's red and smooth. Does it taste sweet?" "What does this pea feel like when you squish it? What color is it? Is it soft?"
2. Try the candy game. I suggest using jelly beans.

LEARNING TO TASTE: CLASS BASICS

<u>The Jelly Bean Game</u>
Works best with ages: 5 and up
Main idea: Talk about a seemingly simple food and see how taste preferences differ.

Preparation

- An assortment of high-quality jelly beans of at least five different flavors. (Jelly Belly tends to have good, flavorful jelly beans, but any brand will do as long as you avoid the really low-end sugary ones.)

Ask your child to choose his top two favorite jelly beans and compare with others. (If you only have one child, this can be a friend or you.) Talk about why he prefers one color and you prefer another. Show how preferences differ with different people. If you have a large enough group of children, let your child see where he stands in relation to the average preference.

Before eating the jelly bean, get your child talking about what it looks like, how it feels and how it smells. Let him or her experiment:

- What's different about the two jelly beans? Color? Smell? Texture?
- Does everything look, smell and feel like you were expecting?
- What happens if you squish it? What changes? Is there a different color on the inside? Does the smell change?
- What is the jelly bean "saying"?
- Which jelly bean do you think will be the sweetest?

Then, let him or her taste the jelly beans, one at a time.

- What does it feel like to have the jelly bean in your mouth? Hard? Soft? Squishy? Grainy? Chewy?
- Did it make any sound? What kind?
- How does it taste?
- Which one tastes sweeter? Better? Why?

Let your children select a few more jelly beans.

- How is it different to eat two jelly beans at a time? Does it feel different? Smell different?
- Can you taste each one individually or do you lose the individual taste? Is the new taste better or worse than the original taste?

If you've bought one of the deluxe Jelly Belly sets where they provide recipes to mix jelly beans and create new flavors, try a combination of different flavors, for example the one that makes root beer or peanut butter and jelly. This is a good way to show your children that tastes mix to create other tastes – but you have to get the mix right.

- What happens when we mix the right tastes?
- Can you taste the original flavors?
- What happens when we get it wrong?

<u>Variation for Ages 3-5</u>
Follow the exercise above, but substitute different flavors of Jello for the Jelly Beans. Make at least five different flavors in small Dixie cups.

Learning to Taste: Class Basics

Class Two: Understanding the four main tastes

In the second class, the focus is on the four main tastes. (The more recently discovered fifth taste, *umami*, similar to the taste of chicken soup, is not included in this class.)

The children fully experience the four main tastes but also learn how their tastes differ from others' tastes, what foods they prefer, and how to use basic ingredients like salt and sugar in cooking.

<u>Exercise One</u>
What are the four tastes and what foods have them? Children name the tastes and brainstorm foods that represent them. For example,
Sweet = candy, cake, carrots, orange juice, ice cream
Salty = salt, potato chips, oysters, ham, salted butter
Sour = lemon, vinegar, green tomatoes, pickles
Bitter = endive, artichokes, grapefruits, olives
Mixes = dark chocolate (bitter, sweet), a very ripe green apple (sour, sweet)

<u>Exercise Two</u>
How do we each experience these tastes?
Each child receives five small glasses with a solution representing each of the four major tastes and a glass of plain water. The teacher asks the children to take a sip of each glass and mark down the taste and also to note if they find the taste "sweet" or "too sweet," "bitter" or "too bitter," etc. The teacher tallies up the class totals and shows how tastes – even of the exact same food – vary among different children.

<u>Exercise Three</u>
What does each child prefer when it comes to food?
The teacher brings out several different foods and asks children to taste them and write down which one they preferred.
Example One: Regular bread versus bread made without salt. This example teaches children about the role of salt in balancing out flavors.
Example Two: "*Un cake salé*" (A savory cake with pieces of meat cooked into it) versus a sweet apple cake. This lets children see whether they prefer sweet or salty foods offered in the same format.

Example Three: Fresh lemon juice mixed with water (prepared in front of the class) versus the same lemon juice with the opportunity to add up to five sugar cubes. Here children experiment with sour and sweet. Who liked the sour taste? Who needed five sugars to make the lemonade sweet enough to drink?

How you can try it at home

1. Talk to your children about the four tastes and let them think of an example of each. Point out to them during meals which tastes can be found in each dish or whether you included salt or sugar in something you prepared.
2. Try the lemonade exercise above. Let children make their own lemonade (with your help) and see how many sugars it takes to make each one of them find the taste appealing.
3. Try the Four Tastes game.

The Four Tastes Game
Works best with ages: 3 and up
Main idea: Help children to recognize and discuss the major tastes

Preparation:

- Representative foods from the four major tastes

  o Lemon (sour)
  o Banana (sweet)
  o Dark chocolate (bitter/sweet)
  o Jam (sweet)
  o Saltines (salty)
  o Grapefruit (bitter)
  o Salted almonds (salty)
  o Tomato (sweet/sour)

Give each child a little of each food, asking him to name both the food and the major taste that's present. (For younger children, you'll probably need to tell them as they taste each food.) Debate whether each child thinks the chocolate tastes sweet or bitter or the tomato tastes sweet or sour. See where the points of difference are and discuss them.

LEARNING TO TASTE: CLASS BASICS

Ask children to brainstorm other ideas of foods that fit into the different taste categories.

Class Three: Preparing a meal

In this class, the children cook a complete meal. Each child gets a different task to complete. One meal often used is an endive and poached egg appetizer followed by roast beef and green salad. Drinks are fresh lemonade or water. They also set a table in the classroom/kitchen and try to make the atmosphere pleasant with pretty table decorations and limited noise.

During the meal, the teacher explains the interplay of flavors in the dishes and asks children to describe the different foods they're experiencing by using the sensory words they learned in Class One. For example, the teacher points out the sweet quality of the beef, allowing children to add salt if they like in order to balance the flavor. Each child tastes both the edges of the meat, which, due to cooking, have become slightly bitter, and the middle, which is more tender and sweet. The smells from these two different parts are different as well. Children are encouraged to chew longer than they are used to chewing to fully appreciate the flavor. The same analysis is applied to each dish. At the end, the children rate how they liked each dish.

How you can try it at home

1. Let your child help out in the kitchen by giving him or her a small role in the overall meal. For example, teach him to make the salad dressing. Not only should this get him involved in cooking, but also it will encourage him to eat his greens! Show him the basics of putting oil and vinegar together with a little salt and pepper and then let him experiment by adding what he likes. Some possibilities are:

   - Different vinegars – red wine, white wine, balsamic…
   - Different oils – nut oils, sunflower oil, olive oil
   - Mustard – whole grain, standard, mixed with various flavors/spices
   - Freshly-squeezed lemon or lime juice
   - Orange (particularly if you then add some orange slices to the salad)
   - Sea salt

LEARNING TO TASTE: CLASS BASICS

- Pepper mixes

Let your child use his creativity to explore different possible combinations while learning which ones taste good. You may have to accept a bit more mess in the kitchen and the occasional inedible concoction, but it should be well worth it!

<u>For Younger Children</u>
Make the salad dressing with them, letting them add different elements, with your supervision and stir. Let them dip their fingers in the different oils or vinegars and taste them.

<u>For Older Children</u>
Let your child become the house specialist in salad dressing. Put her in charge of the salad course and make sure you talk about each creation she comes up with. Building on what she may have learned in science class, you can also point out that oil and vinegar will separate if left too long.

1. As you are cooking the rest of the meal explain to your child why you've matched certain foods together, what you're doing to cook them, how you choose the foods. Talk, talk talk!
2. Work your children's creativity by letting them decorate the table. Ask your child to make something nice to put on the table, perhaps using colored paper or ribbons or candles. Suggest he find something that goes with the color, taste or theme of the meal.
3. Rather than making something, ask your child to be creative for other ways to decorate the table. For example, send him outside to gather a few leaves, branches or pine cones for decoration or let him choose among the various decorative items you have at home: candles, vases, special napkin holders, figurines, etc.

Class Four: Smell

Since smell accounts for such a large proportion of taste, Class Four aims to increase children's sensitivity to smells. Children are asked to smell a number of different little bottles containing different smells. The first set includes flower smells, the second, spices and the third common household smells and others (including pine, olive oil, nut oil, bitter almond, peppermint, tar, lemon liqueur and others). They then try to identify what they are smelling.

Afterwards, these odors are brought back to the world of taste and cooking. Children are asked to compare the smell of a raw carrot with one that's cooked, along with the water it was cooked in to understand how some of the original smell was "transferred" to the cooking water. They compare fresh oranges with marmalade, sparkling water with still, four different syrups (cassis, strawberry, lemon, mint), three cooking oils and two different types of honey. With each new food, they identify it, describe it, express an opinion about it and write down that opinion.

How you can try it at home

1. Take your children on a "smell tour" of your kitchen, letting them smell everything you've got: cooking oils, fruits, vegetables, meat. Let them compare the smells of real foods like fresh fruits with that of pre-packaged ones, such as breakfast cereal. Point out that many "real" foods have enticing, interesting smells while pre-packaged food does not.
2. When you're cooking, let your children smell the foods at different times. If you're making a stew, for example, let them smell the individual ingredients before they go into the pot. Then let them smell them in the pot, with spices added and as the stew cooks. Talk about the differences.
3. Play the Taste and Smell game.

Taste and Smell Game
Works best with ages: 5 and up
Main idea: To show how vital smell is to taste

Preparation

- Blindfolds or something to cover your children's eyes so they can't peak

LEARNING TO TASTE: CLASS BASICS

- Foods with similar textures but different tastes, such as

    o   A ripe apple and a ripe pear
    o   Fruit-flavored baby foods, at least two different kinds
    o   An assortment of different colored jelly beans

Blindfold the children and ask them to hold their noses while they taste each pair of foods. Can they tell the difference between the foods without being able to see or smell them? What do they taste? They probably can only tell the taste (in this case, sweet) and not the overall flavor which includes smell. Let them try to guess what they're eating and explain to them that without smell, flavors are much less interesting.

<u>Variation for Ages 3-5</u>
Place a variety of fruits on the table in front of your child and hold each one up to her nose so she can experience the different smells. Then allow her to choose which one she wants to eat!

Class Five: Sight

What role does sight play in how we enjoy our food? In this class, the teacher talks about vision and how we use it to understand and enjoy our food. She leads the class in a game linking colors with seasons. The class matches pictures of flowers, fruits and vegetables with their names. Afterwards, the children are invited to describe a plate of attractive-looking mashed potatoes in terms of color and appearance. When they taste them, they find that the mashed potatoes are very salty, probably too salty for their liking, thereby learning that appearance is not everything. They also get to taste two cakes: one which doesn't look very appealing, but tastes delicious and the other which looks great, but has only a mediocre taste.

How you can try it at home

1. Make sure to talk about how your food looks before eating it. See the list of food words in Class One to get you going. Get your children to start thinking about how to make food look attractive. Let them decorate dishes before you serve them either by deciding where each food item should go on the plate or by adding chopped up fresh herbs, spices or a sauce.
2. For one meal, make a simple cake and give your child free reign to decorate it. Make sure to have the right supplies on hand: icing and food coloring, decorative candies (jelly beans, sprinkles, M&Ms, skittles, etc.), fresh mint, candied flowers, powdered sugar or anything else you can think of.
3. Play the Taste and Sight game.

Taste and Sight Game
Works best with ages: 5 and up
Main idea: To show the influence of sight and visual assumptions on taste

Preparation:

- Blindfolds or something to cover your children's eyes so they can't peek
- Three different flavored soft drinks

Learning to Taste: Class Basics

- Unflavored seltzer water and food coloring

Prepare four glasses of a small amount of each drink. The first three should contain different fruit-flavored soft drinks and the fourth should be seltzer water with food coloring. (Make the drink a different color than the three original ones.) Ask your children, or have them ask an adult, to taste all four glasses and identify the flavor in each one. See how many think the fourth cup is a soft drink!

Once the trick is revealed, use the first three soft drinks to try a variant on this test. Blindfold the participants and ask them to taste the three soft drinks. Can they identify which one is which without seeing the color?

Discuss with your children how sight can affect taste. Point out that things can actually taste better (such as healthy foods like vegetables) or worse (such as mass-marketed, pre-made foods and drinks) than they seem!

<u>Variation for Ages 3-5</u>

For the younger child, focus on exercises one and two above. Let your child help decorate dishes as you prepare them and ask her advice about which decorative element you should add to the dish – "Should I chop up the basil in pieces or just place the leaves on top of the fish?" "Do the red or blue berries look better on this ice cream?"

Also, for one meal, bake a simple cake and ask your child to decorate it, being as creative as possible. Apart from icing and candies, maybe her favorite plastic toy would look good in the middle? (Just make sure she doesn't eat it!)

Class Six: Touch

How does touch influence what we like and dislike about food? In this class, students brainstorm different textures and then associate foods to each. For example:

Rough = a winter pear
Slippery = soft-boiled eggs
Supple, soft = sponge cake
Creamy = whipped cream
Velvety = creamy soup
Sticky = cotton candy
Flaky = the pastry on an apple turnover
Greasy = butter
Stringy = asparagus
Silky = chocolate mousse

Combinations: Some foods can combine hard and soft like a ripe peach and its pit.

Temperature plays a big role in how foods feel. Children are asked to touch and classify water at different temperatures. They then taste the water and talk about what they've experienced. For example, cold water gives the impression of being thirst-quenching while lukewarm water often does not.

The final part of touch is how foods feel when you eat them. Children experiment with tasting different foods with similar tastes but different textures: a lemon sorbet versus a lemon granita, a hamburger versus a steak, different types of pastries – some denser and some flakier.

How you can try it at home

1. Talk about texture with your children and let them feel their food (even if it means relaxing dinnertime manners a bit!)
2. When you bring home a few bags of groceries, ask them to close their eyes and guess what's inside by feel. Can they tell the difference between oranges and grapefruits? Can they feel the delicate fuzz on a peach? How does a zucchini feel different from a cucumber? Can they recognize asparagus by its tell-tale tip?

LEARNING TO TASTE: CLASS BASICS

3. If you go food shopping with your children, teach them how to assess foods by feel. Is this bread fresh? Is this avocado ripe? Are these tomatoes overripe?
4. Play the Temperature game

Temperature Game
Works best with ages: 5 and up
Main idea: To show children how temperature affects taste

Preparation

- Ice cubes
- A banana or peach
- Apple pie

Ask children to suck on an ice cube and then eat a piece of banana. They should notice that it's much harder to taste the banana when their mouth is cold. You can try the same experiment with a peach, or simply by putting the fruit in the refrigerator for a few hours before starting. Moving on to apple pie, ask your children to try a piece that's cold and compare it to one that's warm. Which one do they prefer?

Variation for Ages 3-5

Focus on texture rather than temperature for the three to five year old. With your child in the kitchen, make both mashed potatoes and boiled potatoes (or, if you prefer, scrambled eggs and hard-boiled eggs). Ask him to feel the potato or egg before you cook it and then afterwards in its two different states. Point out that this is still the same food, but the texture has completely changed!

Class Seven: Taste under Attack

I love the name of this class, "*Le Goût Face à Certains Agressions*" or Taste under Attack. What does it mean? Two things: First, children learn about foods that have "aggressive" tastes. Under foods that "attack" the eater, the teacher lists those with a metallic or astringent taste as a result of poor storage or preparation. Canned vegetables or drinks, for example, or milk that has spent too long in the open air can have unpleasant metallic tastes. Astringent tastes can be found in wines, coffees or cocoas and are difficult for children to accept. Burning sensations like those associated with certain spices can often be pleasant for adults, but rarely taste good to children. Prickly tastes like the feeling of sparkling water are also something often prized by adults and hated by children. Children learn that foods can have strange tastes which seem unpleasant at first but which adults can love.

The second part of Taste under Attack teaches how outside influences can affect the eating experience. A lot of noise takes away from the eating experience, not only because we can't fully appreciate all the sensory information coming from the food itself, but also because outside noise can be a distraction. Children learn that the best way to appreciate a meal is in a calm environment.

Children experiment with foods that have different harsh tastes, including sparkling mineral water, chile peppers and others to understand how vast the world of taste is. They then experiment with taste and environment: Children are asked to take a few sips of lemonade in the quiet of the class and then while wearing a Walkman playing a tape of the loud noises of a school cafeteria at lunch time. They write down their impressions of how the lemonade tastes both times and compare them with the rest of the class.

How you can try it at home

1. Let your child try sparkling water. Before drinking, have your child put his or her face up close to the liquid and feel the bubbles. Most children won't like sparkling water the first time around, but it's a great taste to develop because it simulates the bubbly feeling of soda without the sugar and calories.
2. Play the Noise game

LEARNING TO TASTE: CLASS BASICS

<u>Noise Game</u>
Works best with ages: 5 and up
Main idea: To show children how noise decreases the ability to taste.

Preparation

- One of your child's favorite foods
- A Walkman or radio or another child willing to make a lot of noise (this shouldn't be too hard!)

Sit your child in front of a plate of one of her favorite foods. Then, turn up the noise! Ask her to talk about how the food tasted. If she finds that taste is diminished, don't forget to point out that this same effect can occur during family dinners with the television or radio on or with someone talking too loudly!

<u>Variation for Ages 3-5</u>
Teach by example for the younger child. Make sure he eats in as calm an environment as possible. Let him know that tantrums or excessive noise during dinner take away from everyone's enjoyment in eating.

FRENCH CHILDREN DON'T GET FAT

Class Eight: Regional cuisine

In this class, children learn the culinary differences among the different regions of France. After a brief lesson on differences in climate and agricultural resources in the different regions of France, children begin to understand how food traditions developed differently. They're then tested on the regions, their climate and their foods while they taste specialties from different regions, including fondue from Le Savoie, the cheese Saint-Nectaire from the Auvergne region, and oysters (for those who're willing) from Brittany. Children also smell roses from the Loire Valley, violets from Toulouse and cassis from Burgundy.

How you can try it at home

1. Although French children learn about French food, I suggest you teach your children about food from all over the world since America has so many different cultures and ethnicities. Introduce them – even from a very young age – to all sorts of foreign foods. Check with your doctor, but most babies can happily eat just about anything, even hot foods like curry and spicy salsa. Try to find the authentic stuff (i.e., a giant burrito stuffed with fried chicken is not really very Mexican, and New York style pizza doesn't count as Italian), and make sure you give them the background on what they're eating, even if it's just to tell them where the food comes from and why it became popular there (a little internet research may be necessary here – if your child is old enough, he can do it himself!)

Exotic foods can play a great role both at home and when you go out. At home, they spice up an evening meal and can be fun and interesting for you to make with your children. Roasting the eggplant for babaganoush on your stovetop, for example, is a fascinating food experiment that turns into a great tasting dish. Frying up a bunch of vegetables in a wok can also provide a great show for children and gets them eating healthy foods.

Going out to ethnic restaurants can also be a big treat for your children that makes eating fun and exciting. If you're lucky enough to have an Ethiopian restaurant near you, I highly recommend going with your children. At these restaurants, everyone eats with his hands using a

LEARNING TO TASTE: CLASS BASICS

big piece of doughy bread. A Japanese steakhouse where they cook at your table also makes food preparation into an entertaining night out. Indian, Thai or Vietnamese restaurants with their variety of different spices and dishes will also do the trick.

Also, if you do need to get some quick food on-the-go, think ethnic. Avoid McDonald's and head for a Middle Eastern take-out place where you can get your children a falafel sandwich or some stuffed grape leaves. This way, you'll still have used that particular meal as a learning opportunity for your children.

2. Play the Exotic Foods game

<u>Exotic Foods Game</u>
Works best with ages: 3 and up
Main idea: To introduce your child to the variety of foods that can be found in other cultures and expand his or her repertoire of tastes.

Preparation:

You'll need to collect a range of different foods from different cultures. Choose your favorites, but stay away from the traditional, fatty foreign-American dishes like sweet and sour chicken and tempura. A few suggestions:

- Edamame (Japanese soy beans in the form of peas in the pod)
- Vietnamese summer rolls (the kind that aren't fried)
- Hummous (chick pea paste from the Middle East)
- Camembert (French cheese)
- Chinese steamed buns (doughy buns with fried meat inside)
- Vegetarian sushi (perhaps try oshinko maki, a Japanese pickle inside sushi rice and seaweed, or for a less strong taste, avocado maki, avocado inside sushi rice and seaweed)
- Tamales (cornmeal patties, often with chicken and vegetables, cooked inside corn husks)
- Kimchi (Korean fermented cabbage – many varieties exist, the strongest of which are not for the faint of heart!)
- Tacos (the real Mexican kind with a soft corn flower shell and preferably a green salsa)

Introduce each dish to your child and try it along with him. Once again, ensure that you discuss the dishes using the chart from Lessone One as a guide and explain the background of each food.

LEARNING TO TASTE: CLASS BASICS

Class Nine: Recap

This class summarizes everything the children have learned and runs through some more taste exercises. Here are a few of them:

- Children are given a lemon candy to taste and then asked to hold their noses closed, hence cutting off most of the flavor. They talk about what they taste both before and after. (Main idea: Smell is a large part of flavor.)
- The teacher gives a teaspoon of a strawberry short cake to students and asks them to swallow it without chewing. They then taste another spoonful, this time chewing it, and compare the difference in taste. (Main idea: Chewing releases flavor.)
- Children try a small portion of pastry, both hot and cold, and compare the different sensations. (Main idea: Temperature dramatically changes how something tastes.)
- The teacher puts out other foods for the children to describe, using everything they've learned during the course. The foods include milk products like cheese and yogurt, toast (made from good French bread, of course!) with butter and different types of oils, sausages from different regions in France and four types of exotic jams. The children are then free to prepare a piece of toast for themselves with their choice of all the foods served. The class discusses the different preferences of the group. (Main idea: Everyone has his or her own preferences where food is concerned and it's important to know what you like.)

How you can try it at home

Try the games used in Class Nine at home or some of the others that you'll find at the end of this book.

Class Ten: Top Notch Gastronomy and Closing Ceremony

For the final class, a professional chef comes to prepare and share a relatively sophisticated meal with the students. The students are invited to comment on and discuss the food preparation and the taste of the food.

That's it! A quick graduation ceremony where each student gets a taste certificate and they head out, ready to taste the world!

How you can try it at home

1. Unless you happen to be a professional chef or know one, I'd suggest taking your children of any age out for good, professionally cooked food in a nice restaurant. Find a nice place, I'd recommend a French restaurant, where you can try healthy foods cooked in ways that you simply don't have time to do at home. Make a reservation for sometime early so the restaurant will be relatively quiet. (You might even be able to take advantage of an Early Bird special or a *prix fixe* menu - a menu where everything is included for what is normally a reasonable price)

Suggest to your child that she try something different or adventurous or a food that she doesn't particularly like at home. (Avoid at all cost her ordering the lone hamburger that may be on the menu.) Share all your meals among the family and comment on each dish, asking your child to use his newfound vocabulary to describe the dishes and discuss how they appeal to him or her.

2. Once you've run through a few taste classes with your child, let him feel like a taste expert. Making a certificate of achievement is a great idea, and you can even present it in a little ceremony at dinnertime. But, the education shouldn't end there. Now that you've helped your children to be able to think and talk about what they're eating, make sure to include them when you discuss a dish. This will reinforce what they're just beginning to understand and help them to make healthy and intelligent food choices as adults.

Beyond the Classroom: Teaching Taste at Home

Ok, so you've tried some of the exercises. Now what? Taste classes are a great introduction to thinking about food, but they work best in combination with an environment at home which promotes healthy eating. In France, families have a range of eating practices which do just that. In order to change your children's views on food (and maybe even your own), you'll need to integrate several elements of this method into your lifestyle. Don't worry! Most of these changes are simple and require only small modifications to your daily routine.

Aren't taste classes enough? No. Remember, the "do as I say, not as I do" principle: Children learn how to behave by observing their parents. It doesn't take a child psychologist to tell you that if you try to force steamed broccoli on your child while you eat pizza, it's not going to work. The same goes for your everyday habits. If your child sees you constantly snacking or binging on junk food, chances are she will grow up doing the same, regardless of whether you educate her to like good food. (So by following this method you may even wind up losing a few pounds yourself!)

Before we start, I have to tell you one thing. Learning to appreciate food and tastes is hard if you have a diet of frozen or pre-prepared foods. One major reason for the French ability to eat good, healthy food in small quantities is that their food is often natural, homemade and just plain tastes good. Yes, in order to allow your children to enjoy a rich variety of good-tasting foods, you're going to have to do some cooking. Don't despair! Cooking does not mean spending hours planning meals, buying groceries and preparing that perfect sauce. If you're not already in the habit of cooking, buy a quick-meals cookbook or find a good website that gives quick recipes. Chicken with a few sautéed mushrooms and green beans can take ten minutes to make. Adding some fresh, chopped flat-leaf parsley or another favorite herb makes the dish pretty and better-tasting and takes only another minute.

Those of you who already cook on a regular basis may need to learn to diversify some of the foods you buy and prepare. If you find yourself buying the same things from your local supermarket week after week, it's no wonder you and your children are eating too much or looking for satisfaction from extravagant desserts. We eat too much when our taste buds aren't properly stimulated. While there's no need to give up your family's favorite foods, you should also start incorporating something new, at least on a weekly basis. You can follow through on the taste lessons by talking about each new food item in terms of its taste sensation, its history, and all the other aspects your children learned in their taste lessons.

Here are a few basic rules to follow with your children that should help you to create the right food environment at home.

<u>Eat everything</u>

Show your children, by example, that you enjoy a wide range of foods. Studies have shown that the sight of a mother eating a particular food can encourage a child to try it.

As I mentioned before, the range of French food is vast. Even just the range of vegetables is enormous. French families try to eat different things all the time and they benefit not only from better nutrition through variety but also from rarely having to hear their children say "Broccoli <u>again</u>, Mom?" Even if broccoli is being served for the second time in a week, it's no doubt being prepared in a very different way, which also means that even if the children didn't like it the first time, they may like it this time.

You can start by reconsidering the foods in the supermarket or your local grocery store that you normally, often unconsciously, pass by. How about endive? The bitterness makes it an acquired taste, but sautéed in olive oil with just a touch of sugar, it turns into a sweet delight. What about a type of fish you don't normally buy – or any fish at all if you normally eat meat? How about couscous to go with a quick Moroccan-style stew? Often, these new foods are no more difficult or time-consuming to make than your old standbys, and they have the huge plus of helping your family to maintain a healthy weight and enjoy good nutrition.

Take something as simple as apples. Maybe your child likes them, maybe she doesn't. If she doesn't, this could be because she's tasted a few varieties, but has not experimented with the wide range of colors and flavors classified under this same category: Granny Smith, Golden Delicious, Rome Beauty, etc. Maybe she doesn't want to eat the raw apple sitting in your fruit bowl, but she'd happily eat apple pie, applesauce with cinnamon or baked apples. The same goes for any fruit or vegetable that you want your child to appreciate!

Do a little research on the internet for recipes or buy a French cookbook for some ideas on new foods to try and ways to prepare them. Even the old favorites can be turned into something more appealing and exciting if you find a new recipe. Try to make mostly relatively quick, simple meals. If you like to cook, you may be tempted once in awhile to make a special something that takes the better part of a weekend to finish, but this will never become something you can reliably cook for your family on a regular basis. Who has the time? The secret of French cooking is to have a number of old stand-bys - good-tasting and nutritional dishes that are easy to make, along with the creativity to modify them slightly with a different seasoning or presentation.

Variety is not only the spice of life, it is the spice of any meal! If one food isn't working for your child, try a different one. Since our senses adapt to a food and the taste diminishes as we eat it, it's better to switch foods a few times during a meal in order to keep experiencing the full flavors of a particular food. Coming back to the original food, you'll find that the flavor is back in force. This explains why, "full" after eating the main meal, parents and children alike still "find room" for dessert. It's simply that we're tired of eating a food that has become relatively tasteless and are even more attracted by a new, presumably appealing taste. You can apply the same logic to other, non-dessert dishes. Taking a few bites of meat, for example, can invigorate the senses to come back to the vegetable dish. I've known friends whose children insisted on eating all of the rice before they would even take a bite of the meat or vegetable on their plates. You can overcome this problem by making a game out of "eating around the dish", taking three or four bites of each food on your plate, and by setting the example for your child by doing it yourself.

Another reason why French meals work so well is that the French often eat many courses, each with small portions. This way flavor is maximized; as soon as our taste buds get maxed out on a certain food, we go on to the next course. This method also allows you to enjoy eating foods that may be a bit more fattening. Eating in small portions blunts the effect of the higher calorie content, and yet it also satisfies you.

<u>Eat together</u>

Make eating together a pleasurable, family experience. Ok, I know this is hard to accomplish. With the children's many after-school activities and your own hectic work schedule, it's not something that today's American family can do every day - but try to eat as a family as often as possible. It's worth the extra effort. The more structured your meal times are, the more likely your children are to associate food with sit-down meals and therefore avoid snacking, "unconscious eating" and fast food. This last point is a key reason why the French eat well – they rarely snack. Consuming food without the social ambience that goes along with it is only half the fun, and the French know it. Eighty-eight percent of them think that a meal – any meal - should be shared with others. Try also to avoid arguments and stress and focus on the simple pleasure of eating good food together. A meal is not the time to talk about a bad grade or a family problem. For the French, eating equals a pleasurable time spent in the company of others. If you have a problem to discuss, do it after dinner.

Another benefit of eating as a family is that everyone, including children and other picky eaters, is more likely to accept new foods in social settings. A survey by anthropologist Jane Kauer asked people to imagine themselves in three different scenarios, each time confronted with a food they hated. In the first situation, they hadn't eaten for 12 hours, in the next, they hadn't eaten for three days, and in the final situation, they find themselves as a guest at someone else's house and the hated food is part of the meal being served. Surprisingly, in spite of their obvious hunger in the first two scenarios, it was the last situation that inspired the most people to eat the hated food. While it is true that older children will be more sensitive to the social etiquette surrounding a meal, even younger children may choose not to refuse a food if they're with others, particularly if the others are children older than they are who are happily finishing their plates. When my two-year-old son Max is eating around a table with his cousins, particularly if the beautiful, four-year-old Mathilde is present, he's much more likely to eat what he's served.

In France, it's simply not considered socially acceptable to refuse a food served to you at a dinner party. It's expected that everyone will be willing to eat all the foods served. (The only exception is the cheese course which can be politely refused, safe in the knowledge that it was most likely not prepared by your hosts.) The French are also constantly inviting friends and family over for dinner, often at the last minute. (While this custom provided me with some challenging last-minute meals as a new wife in France, it also offered a lot of insight into how the French eat.) I often think that this emphasis on socializing contributes to their healthy attitudes about eating, since we're much more likely to try new things and eat reasonable portions in social settings. My worst experiences with binge eating are always when I'm alone and in front of the television where it's easy to eat an entire pizza without a second thought!

Americans tend to think of dinner parties as highly-structured affairs where people are invited a month in advance and the food is carefully thought-out and prepared over the course of several days. By all means, if this is something you enjoy, do it, but don't think this is the only way to have company at dinner. Obviously, it's not practical to expect to have guests all the time. The idea of company for dinner, even close friends and family, can be stressful. But it's something to try to aim for as much as you can. Invite a family friend or two on a regular basis. Not only will the presence of a guest encourage everyone to avoid unpleasant topics, but it also will create a more festive, fun atmosphere that will make your children more likely to eat their veggies and try new things. Making dinner for a few extra people is not very different if you're already cooking, and the added cost will be recouped when they invite you back in turn!

Eat "Real" Food

Just as you wouldn't get very far teaching your children about music using only eighties pop tunes, you can't teach your children to taste with pre-packaged convenience foods. The French certainly don't. They eat real food, often bought in markets, local specialty stores or well-stocked supermarkets, and prepared at home. Most French women I know in Paris would not think twice about heading to the other side of the city to a specialty market if it meant getting exactly the type of chicken they wanted to cook that evening. Every French food store worker is capable of recommending which products are the freshest that day along with recipes that might enhance the flavors of any of their products.

Not so your average American supermarket or dinner menu! Frozen lasagna, packaged bread, cookies or other foods prepared who knows how long ago are bound to have lost their flavor. Ideas about good, fresh food were once a part of our culture. Ralph Waldo Emerson famously said, "There are only ten minutes in the life of a pear when it is perfect to eat." Ever since the processed food revolution of the 20th century, however, fresh has been replaced by convenient as the most important quality in food.

Now, not everyone has the time to drive 20 minutes out of her way to the local farm store or market, so improvise. If supermarkets are your closest, most convenient way to shop, try to follow these few simple rules:

- Look for the freshest fruits and vegetables. Try to choose what's in season and don't buy the same ones every week.
- Buy at the meat or fish counters rather than in the pre-packaged section.
- The benefit of a supermarket is the wide variety of foods available: Reconsider foods that could easily make a tasty hors d'oeuvre or add zest to a dish but which you don't usually buy like radishes, different types of nuts, olives or avocados.
- Buy fresh herbs when you can.
- Bring recipes or ingredient lists with you to make sure you get what you need and won't have to resort to ordering in pizza at the end of the week because you've forgotten a key ingredient and don't have the time to dash out to pick it up.
- Avoid most of the pre-packaged aisles if you can, particularly if you're with your children: Looking at the boxes of cookies and cakes will only increase their (and your) desire to buy them.

Since all children like foods that are familiar and dislike ones that aren't, you'll be starting your children off right if you get them familiar with fresh, healthy foods. Talking to my son's babysitter the other day, I saw yet another example of this. She watches both my son Max and a little French boy named Natal. Natal's mother, the counter-example to the traditional French mother I am describing here, brings over mostly pre-packaged baby meals. Our babysitter tells me that every time she feeds Max his home-cooked meals, Natal shouts that he wants some - and vice-versa. However, one bite of the home-cooked stuff and Natal makes a face and spits it out, as does Max when he gets a taste of the processed food. The lesson? Both like only what they are used to, but in this case, one is used to freshly prepared meals and the other to factory-processed ones.

Part of a "real foods" strategy is understanding what your child eats when he is not with you. Institutional meals can be dreadful, whatever the budget of the institution. In France, many schools and daycare centers make their weekly menus available to parents and I've heard of

BEYOND THE CLASSROOM: TEACHING TASTE AT HOME

cases of parents lobbying for better or more diverse food. Make sure you're informed about what's being served to your child. Even if you can't change it, you can be sure to balance out what they're eating by what you serve them in the evening. For example, if they've eaten chicken fingers, french fries with ketchup and chocolate cake for lunch, ensure they have a homemade vegetable soup for dinner and a yogurt for dessert. Or, since many schools publish their lunch menus, try to pack a lunch for your child on a day when the food at school is unacceptable.

Ideally, school meals should be reasonably balanced, helping children learn the correct way to eat. They should also be pleasant, lasting at least 30 minutes in a (relatively) calm setting. If you have any influence over this, more power to you. If not, make sure you reinforce at dinner that meals are calm, fun times.

Try to steer caregivers and grandparents towards a similar way of thinking, even if it means preparing foods for your children to take with them. If they're exposed to bad foods and bad eating examples though, perhaps at a friend's house, simply compensate with their meals at home rather than scolding or stressing about it. Remember, anxiety about food is easily passed on to children! Also remember that the body is forgiving, and even if your child eats something that isn't nutritionally sound once in a while, the key to good weight control and good health is in the larger, overall picture.

<u>Pay attention to what you're eating</u>

No dinners with the television on! A recent study showed that we're much more likely to eat more food when we're not paying attention to it. Anyone who's eaten a bag of potato chips in 30 seconds flat during the scary part of a movie can confirm this. Also, it's hard to appreciate flavors when you're focused on something else. Mindless eating is what causes us problems. Fully appreciating your food means taking time to consider its impact on your five senses and sharing the experience with others. Think about what you're tasting and talk about it with your family. Ask them if they like it, or how they would suggest it be prepared next time. (And be prepared for some "smart" answers like "Never!" Remember that it's hard to change patterns and attitudes overnight, and you as the parent have to remain patient. Just keep trying.)

Eat in a calm, pleasant environment to enable your family to pay full attention to the meal. Try to minimize outside noises and disturbances, odors from the kitchen or elsewhere, or any kind of visual distraction. French people are fanatics about escaping kitchen smells during dinner (which is not always easy in tiny Parisian apartments!) Since taste is 90% smell, you simply cannot fully experience your food if you're smelling something else at the same time. Before I moved to Paris, I used to chuckle at a relative who'd spent time in Paris and who was very choosy about where she was seated in a restaurant. She once even asked to be moved from a certain table because the waitperson was polishing the glass top of the table next to her with Windex. She said the odor bothered her. Now I see that she was right!

Talk to your children about food

Teach your children to talk about their food and the eating experience, not just during your taste classes but all the time. A great way to start doing this is to talk to your children about food yourself. Let them know why you choose the foods you choose, how you cooked them and what they should taste like. Talk your children through their meals and ask them to comment on what they're eating, using each of the five senses. Within the family, find points of similarity and difference in preferences. Don't let them just say, "I like it," or "I don't like it." Instead, have them use descriptive sentences such as, "The smell reminds me of summer at the beach" or "The taste is too bitter." Ok, so children are not always going to be so eloquent, but at least you can put them on the right track by asking questions:

- How would you describe this apple sauce just by looking at it?
- What color is it?
- Is it soft or hard? tender? sticky? firm? crisp? rough? smooth?
- What does it smell like?
- What sound does it make when you eat it?
- What does it taste like? If it's sweet, is it also another flavor?
- What does it feel like on your tongue?

Try to steer your children away from easy observations and into harder ones. (This technique has the added benefit, by the way, of expanding your child's vocabulary.) If words don't come to mind for certain smells, sounds or textures, make them up!

And then more generally:

BEYOND THE CLASSROOM: TEACHING TASTE AT HOME

- Do you prefer your green beans like this or made the way we had them last week?
- Which do you like better, spinach or zucchini?
- Guess which ingredients went into this dish.
- What kind of animal does this meat come from?
- What country does this food come from?
- What's the best season to eat strawberries?

I recently went to a talk by a child psychologist who offered the technique of giving choices as a way to motivate children to do what you want or what's good for them. This works great with food. Of course, you as the savvy parent will stack the choices in your favor rather than leaving them open-ended. Do you want salmon or tuna for dinner? Would you like the mushrooms in your salad or cooked?

If this method seems a bit sneaky to you, remember that children like what's familiar and it takes five to ten experiences with a new food to make it familiar. For many foods, familiarity is the only way forward and the only way to make children familiar with a food is to get them to try it. Once you're over that five-to-ten-experience hurdle, you might be creating a favorite food for life!

<u>Cook with your children.</u>
Turning a bag of groceries into an evening meal is a magical process and letting children share in it will both increase their respect for homemade food and provide them with invaluable information for when, as grown-ups, they'll need to cook for themselves. What's more, the more children help with meals, the more likely they'll be to want to eat them. Nothing helps give children a positive image about food more than seeing how it's produced, prepared and served. Children will feel proud of what they've had a hand in creating and proud that the family is going to eat it.

You don't have to be an excellent cook or have a lot of time to involve children in simple meals. An easy way to involve a child is to let him make a small part of the meal, like making the salad dressing, making freshly-squeezed orange juice to go with the meal, mashing the potatoes or choosing an herb to add to a dish before serving.

Talk to your children while you're cooking, explaining what you're doing and why. Let them taste things along the way and help out wherever they can. This is how both boys and girls will pick up the basics of cooking as well as an appreciation for homemade food. Whereas my French husband is a natural in the kitchen, I am, or rather, used to be, a disaster. Having grown up being expected to help out in the kitchen, he knows how to make a variety of quick, healthy dishes as well as how to save a sauce or meal from impending disaster. I, on the other hand, could mostly be found watching television until my mother called us in for dinner; living on my own for the first time in Boston after college, I found I was unable to make anything more extravagant than pasta and scrambled eggs. It was not until I got married that I learned how to cook, initially by watching my husband. Cooking is a fact of life, whether you're a man or a woman. Why not ensure that it is another skill your children pick up when they are young?

Even when your child eats something pre-prepared that tastes good and is reasonably healthy, this is not necessarily going to help her develop her curiosity about food, including how it is prepared and how its taste can change. Of course, no one can cook absolutely every night, so the important thing is to involve your children as much as you can on the nights that you do cook.

Another great way to get children to appreciate food is to have a small garden in your backyard, or even just an herb plant on your windowsill. Children are more likely to eat something when they've had a hand in growing it and you can show them first hand that freshly-grown foods have a real taste advantage over their store-bought equivalents.

<u>Don't talk to children about diets or calories</u>

By ten years old, I remember already counting calories. A like-minded friend shared with me one of her diet tricks which was to picture chocolate as "something really icky" so she wouldn't be tempted to eat it. (Needless to say, it didn't work.) The Department of Agriculture's Food Pyramid or Food Group campaigns that we learned in school with serving sizes and numbers also left me cold. How could I possibly know what the right serving size was or count how many servings of carbohydrates or milk products I had had in a day? (That said, I was still sad to read that, unable to cope with pressure from food producers' lobby groups, the Department of Agriculture has replaced the Food Pyramid with a mostly meaningless rainbow-colored, pyramid-shaped logo with no food on it at all!)

BEYOND THE CLASSROOM: TEACHING TASTE AT HOME

While it is true that understanding nutrition is important for adults, children cannot understand the link between nutrition and being healthy or feeling good. (I myself discovered it by a process of elimination after a week of pizza and frozen yogurt at college, as I wondered why I didn't feel so well and had no energy, but I wouldn't recommend this method!) Children also can't understand the concept of calories and, except in cases where it's recommended by the doctor, shouldn't be bothered with worrying about eating too many or the wrong type. It's your job to worry about the overall nutritional balance of the meals your child eats. Children should just focus on trying new things and developing their ability to taste. By the same token, don't get your children into the habit of eating low-fat foods. These foods always taste worse than the real thing and there's no reason that most healthy children can't eat a normal-sized portion of the real thing.

Children like to experiment, not to be told what to do. Rather than talking about nutrition, let them understand that it's important and fun to taste and eat a variety of foods, and help them to experiment. Teach them to find the things they like among a range of nutritious foods and to be open to everything in order to do so. You want children to eat vegetables, for example, but let them try a wide range in order to find their favorites.

Of course they'll love chocolate and candies, but show them that these are foods we eat on special occasions when we can really appreciate the taste. What's more, help them to appreciate real chocolate, the dark, bitter kind that provides a real sense of satisfaction, rather than the tasteless artificial milk chocolate that most American and British companies produce; you can easily eat a bar or two of that chocolate and still feel unsatisfied. If you do have a huge box of jelly beans in the house and you're going to give some to your children, talk to them about which flavors they prefer and why. Get them thinking about the sensation of eating a jelly bean or of combining a few to make a different taste. Even if you're eating candies, there's no excuse for mindless eating!

In France, no one talks about nutrition, yet everyone eats relatively healthily. In America, everyone knows that foods that are good for you don't taste good. That famous phrase, "It's good for you," should be avoided at all cost. No child was ever enticed to eat something because it was good for him. That's adult reasoning, and it will come soon enough.

In any case, a person with an ability to appreciate taste should have no problem eating healthier foods.

Don't be afraid to make food that looks and tastes good

The corollary to not talking to children about nutrition or calories is not worrying about them so much yourself. The favorite food of my sister-in-law's five children is *quiche aux épinards*, spinach quiche. While it's true that quiche has eggs and cream in a pastry shell, and it might not be ideal as a regular standard for an adult trying to lose weight, it's a great introduction to spinach (or whatever vegetable you want to put in it) for your children. The results are clear: My sister-in-law's children actually like and ask for spinach, and also cauliflower (due to her famous cauliflower gratin) and endive (in salad), etc., and they are open to trying other vegetables since they already know and like so many of them. This is a virtuous cycle that many American children miss out on since, viewing vegetables as the pinnacle in healthy eating, we often prepare them steamed or boiled with no sauce, which may make them taste good to adults but can be pure torture for children. Remember, eating foods that taste good, even if they have a few more calories, will satisfy your children and keep them from reaching for the cookie jar two hours later.

In fact, spinach is now one of my favorite foods, even though I used to hate it as a child. It's one of those many vegetables that, prepared poorly, is just awful, but well-prepared is marvelous. Think of the lukewarm, stringy spinach you've no doubt had in a school cafeteria and you won't be surprised to see why children naturally don't like it. Next to that image, however, I put the first time I had Italian-cooked spinach in a homey restaurant near Milan. It was fantastic! Served piping hot with just enough olive oil and garlic to bring out the flavor, it was the ideal accompaniment to fish.

Make sure you don't fall into the "My child doesn't eat vegetables" trap. I've heard a number of American mothers say this, a claim that sounds absurd to a French mother. Ok, perhaps a few of them have a child who actually cannot biologically handle eating vegetables, but the truth is that no child starts out wanting to eat vegetables, yet most can learn to like them. French children do, and this has nothing to do with their genes. In France, studies show that the proportion of children who refuse to eat a certain vegetable is always lower than the proportion who is willing to eat it. Although all of the techniques I talk about in this book should help to get your children to enjoy a wide range of foods, the most

BEYOND THE CLASSROOM: TEACHING TASTE AT HOME

basic one is simply to make foods that taste good, even if it means adding a few more calories.

Vegetable soups are also great for children because they're often easier to eat than the vegetable on its own, while still retaining a lot of the vegetable's flavor. Once again, don't be afraid to add a little cream or to sauté the vegetables before boiling them in the soup stock. Relative to the proportions of the soup, the small number of added calories will hardly make a difference.

Experiment with recipes until you find what tastes good to you and your children. Cooking is experiential. You may need to make the same dish a few times to get it to the point where it really tastes good. Given how people's tastes differ, there's nothing that says a recipe from a famous chef or author is going to taste good to you.

Even little children should participate in the family eating experience. Children's food should not be tasteless. Don't feed your baby bland baby food unless you really don't have the time to make something more tasty. Those pre-packaged baby foods all taste the same! Instead of the labels proclaiming salmon, spinach and potatoes or ham and peas, they could easily be labeled "fish mush" or "meat mush." When you do need to use pre-made baby food, favor the single-flavor variety. Mixing foods all together usually makes them lose their taste entirely.

We often have a tendency to want to feed our children bland foods since we believe that certain foods can be too harsh for their delicate palates. Although you need to watch what you feed your child very early on, from about the age of two, they're ready to try many of the main family foods. After all, children were exposed to all sorts of flavors when they were in the womb; why deprive them now? By the same token, don't get suckered into buying your older children so-called "kids' foods," which normally include such nutritional disasters as chicken fingers or mini-burgers. Most kids can and should eat the good foods that their parents are eating!

Another technique recommended occasionally in France is to give dishes names, particularly names that appeal to children. If your child has helped cook a certain dish, or added something special to it, you can always name it after him. "Brian's chicken" or "Zucchini soup à la Clara" will ensure that your child almost always agrees to eat this food. If you feel like being creative, you can name all sorts of dishes, hopefully adding to their allure: Halloween Soup for pumpkin soup, Pennies from Heaven for carrots sautéed with orange juice and cinnamon, or Smiley Faces for sunny-side up eggs with pieces of sausage that form a face.

Something else the French excel at is making food look attractive and enticing. My mother-in-law always set aside time to decorate a plate, perhaps by cutting up vegetables and herbs and sprinkling them over the main course, or by ensuring that each piece of pie had a sprig of mint on it and a well-placed scoop of ice cream. Children are also susceptible to food that looks good. They, perhaps even more than adults, can appreciate whether a dish is "pretty" or not. Children can also help with the decoration, unleashing their creativity and involving them even more in the meal.

Good-looking food should also be served in a good-looking environment. Buy and use an inexpensive everyday tablecloth or set of placemats; today you can easily find something that doesn't need to be washed and ironed. Find something that looks nice but that you won't worry too much about getting spills on. Make sure that your plates and glasses are also visually appealing. With the range of low-price, colorful dishes at places like Target and Walmart, there's no reason not to have everyday dishes that are appealing. You could even pick up a second set just for a change a few times a week.

Limit the sugar

Americans consume twice as much sugar per person as the French. The US Department of Agriculture reports that we gorge ourselves on 158 pounds of sugar every year, the equivalent of 53 teaspoons per day, and this number is on the rise. The problem is that sugar is commonly added to pre-made foods to make up for the lack of taste. Take salad dressing, for instance. In France, salad dressing is something whipped up with oil and vinegar. In the US, it can be a thick, creamy concoction of honey and mustard, loaded with sugar and artificial flavors and bought in the supermarket. Even seemingly healthy foods like fruit yogurt can be loaded with sugar, some providing over 70% of the daily recommended allowance for sugar.

The real killer for your kids, though, is soft drinks. In France, children drink water with their meals and they have juice on special occasions. In the US, soda is everywhere. It's the largest source of all this sugar we consume – and it's deadly. A Harvard professor found that for every soft drink a child drinks a day, his or her chances of becoming obese increased by 50%.

BEYOND THE CLASSROOM: TEACHING TASTE AT HOME

Sugar dulls our ability to taste, making us unable to appreciate our food and leaving us unsatisfied and ready to eat even larger quantities to get that elusive feeling of satisfaction. You need to steer your children away from it, particularly since it's omnipresent in pre-packaged American foods. Here are a few simple rules to follow:

- No sodas at all, or at least not before or during dinner where they can interfere with the tastes of the meal. (This may mean that you and your partner have to go without too if one or both of you is in the habit of drinking a lot of soda. If this seems too hard, be reasonable about it. Have soda when you eat pizza, but try a flavored sparkling water or a glass of wine with a more complex dish.)
- Go light on the juice. Most juices contain tons of sugar, yet we give them to our children as if they were something healthy. One way around this is to make your own juices using a juicer at home if you have the time. If not, let your child eat the actual fruit with a glass of milk or water.
- Always make your own salad dressing. (Homemade salad dressing keeps well, just like the pre-made stuff, so you don't have to remake it for every meal.)
- Avoid buying pre-packaged sauces, like barbeque sauces, which are loaded with sugar. For sauces that are too hard to give up, like ketchup, try to introduce alternatives. French fries taste just as good with vinegar (an English tradition that made its way to France) and hamburgers with mustard (the French way).
- No sugary cereals in the morning. Children can eat and enjoy adult cereals, perhaps mixed with some plain yogurt. If not, they can have bread or another breakfast food.
- Check the labels on all pre-packaged foods to avoid high sugar contents.
- Make and eat great-tasting desserts, but avoid ones that get all their taste from sugar
- Don't fall into the trap of using sugar substitutes or low-sugar ingredients. These are even more capable of dulling our sense of taste. What you want your children to get used to are more complex tastes that don't require as much sugar, not simpler, sweeter tastes that contain loads of chemicals!

- Use honey instead of sugar where possible since its taste is more refined. Feel free to experiment with different types of honey (there are hundreds!) until you find the one that is most satisfying for your family.

<u>Don't force!</u>
Forcing children to finish their plates can be destructive to their future eating habits. It's been shown that obese children are more likely to have mothers who forced food on them. Children know when they've had enough and they simply stop eating. Free from many of the food neuroses that plague adults, children regulate their appetites much more efficiently than adults do. Studies have shown, for example, that when both children and adults are offered some pre-dinner hors d'oeuvres, children go on to eat less at the main meal while adults go ahead and eat the same standard portion. That means that children eat until they are full and then stop, whereas adults eat what is socially acceptable or intellectually appealing.

Children's appetites can also vary dramatically day-to-day, much more so than adults'. Getting them used to the idea of stuffing themselves for the greater good of "starving children in Africa" is only going to teach them that overeating is ok, even well beyond the point where they can appreciate taste. Serve children very small portions to begin with and let them ask for seconds if they want. Not only does this minimize eventual waste, but also it gives children time to digest and their stomachs time to send the message to their brains that they're full.

Forcing food on children more often than not will make them hate the food and begin to view mealtimes as battlegrounds. Adding this element of struggle to eating tends to diminish if not remove entirely all pleasure from eating. Any child forced to eat green beans will quickly understand that green beans are something horrible that one could not possibly want to eat of one's own volition. I remember my sister forcing me to eat a piece of blue cheese when I was six or seven years old to see what my reaction would be. Because it's an acquired taste, I hated it on the first try, and the memory of being forced to eat it kept me away from cheese for the next twenty years (a severe handicap in France!).

Try also to avoid bribing or negotiating with your children to get them to eat an offending vegetable or meat. By telling a child "Eat your Brussels sprouts or you won't get any ice cream," you let her know very clearly that Brussels sprouts are a punishment she has to endure to get to the good stuff. Your children are smart – they are not going to eat up all those veggies when it's clear that even you don't like them! You also risk creating a power struggle between you and your child that takes away from the focus on taste. In any event, by forcing, bribing and negotiating, you'll never create a long-lasting appreciation of the food being pushed.

By now you may be thinking that I recommend letting children eat whatever they want. Not so! There's a fine line between being firm but not autocratic. Children also need to understand that the meal in front of them is what the family is eating and they don't have to partake if they don't like, but if they won't at least try one of the foods being served, they will not go directly to dessert. A normal, healthy child can afford to miss one or two meals until this point is hammered home. (If in doubt, ask your pediatrician.) This is where the other methods come in: Getting children to suggest what they like and dislike about the food, involving them in the preparation and developing their sense of taste should have them eating more of your vegetable dishes in no time.

By the same token, don't fall into another common trap of allowing children to choose what they eat. Children should not choose their own foods on a regular basis. They should not be allowed to open the refrigerator themselves and take out what they want. The French know this well. My mother-in-law recalls how shocked she was when, while she and her family were living in the United States, her children's friends would come over and head straight for the refrigerator to find something to eat! There are two reasons for this: First, children don't know about nutrition and their palates need to be educated before they can really choose a meal more sophisticated and healthy than macaroni and cheese. Secondly, allowing children to make adult decisions can be stressful for them.

<u>Don't forbid</u>
Nothing helps develop food neuroses faster than being denied (or denying yourself) even a small taste of a food you really want. The French don't deny themselves anything, they just make sure to eat smaller portions of the high-calorie food and only occasionally. We've all known what it feels like to crave a "no-no" food, say a chocolate chip cookie, and to try to avoid eating it. You substitute other foods and when you find that they don't satisfy, you eat something else, and so on and so on. The result, of course, is that you've ended up eating more calories without satisfaction, and you realize that if you had just eaten the cookie

in the first place, you would have taken care of your craving and been able to go on with your day.

For children, the effect is even worse, particularly if they see you eating the food yourself. Later on, when, at a friend's house or elsewhere, they are finally allowed to have some, they'll most likely gorge themselves to make up for the sense of deprivation that they felt at home. Instead of forbidding high-calorie foods, manage them:

- Buy or make the high calorie-food only occasionally. If you have it sitting around, it'll be a constant battle to keep your children from eating it!
- Don't buy pre-processed foods, like boxes of cookies or chocolate bars. These are less satisfying which makes it easier to eat way too many. Plus, you want your child to think of a real treat as a freshly-baked muffin or cookie, not a box of Oreos!
- Present a wider and healthier range of foods as treats, such as honey on bread, yoghurt with jam, a banana with sugar and lemon, sorbet, etc.
- Serve or buy small portions.
- Avoid the guilt! If you've decided to give your children a treat, let them savor it without guilt or stress.

<u>Be honest about tastes</u>

The first time I gave a dinner party to a group of French couples, I was almost in tears. I had foregone sleep to make a variety of my best dishes, mostly American or American-inspired foods, like my "famous" butternut squash soup. Not one of my guests complimented me on the food! Every time I gave a dinner party in an Anglo-Saxon country, and, for that matter, every time I attended one, the compliments, real or fake, abounded almost immediately after the first bite was swallowed. Not so in France! The dishes were instead discussed in detail – was that cinnamon in the soup? Hadn't they eaten something similar at a restaurant in London?

The French are honest about how their food tastes. If it's good, they revel in talking about what makes it good and what might make it better. (The French are food perfectionists, even the ones who don't cook.) If it's not so good, they discuss what's wrong with it and what not to do next time. The subject is complex and cannot be simply summed up as "Great mashed potatoes, Marisa."

What's more, they refuse to accept food that doesn't taste good. This is obvious to any casual American traveler who wanders into an unprepossessing bistro to order a standard French meal: The average food standards in France are higher than in any other country in the world.

French children are raised with these same standards. Even when food doesn't taste as good as it could, they are still encouraged to eat at least a little of it (and the rest of the family does the same) since wasting food is not an option, but they understand that this particular dish was not *'une réussite'* (a success) and it's not necessarily the fault of the ingredients. Next time it will taste better or be cooked in a different way.

Obviously, not everything you cook is going to work out, but there's no question of throwing out a whole meal or making only three or four staples that you can't mess up like hamburgers or meatballs and spaghetti. Instead, if you've prepared a zucchini soup and it didn't turn out exactly as you wanted to, acknowledge it, eat it anyway, and improve it for the next time. Don't let your children go away thinking that zucchini soup is disgusting. Depending on the taste, try adding a little cream, lemon or salt and pepper, staples that you should always have on hand and see if you can change the taste. If that doesn't work, just try something a little different the next time you make it.

No eating between meals

Yes, this sounds more like the advice you would receive in a standard diet book rather than in a book about helping your children appreciate taste. However, there are real taste benefits to not eating between meals. First of all, food tastes better when you wait for it. The long awaited dinner that finally arrives after a hard day when you skipped lunch always tastes a notch better than the same dinner when you're less hungry or not hungry at all. Check with your doctor about how often your child should be eating and stick to that schedule. According to common French wisdom, spacing out meals is also said to be better for the digestion.

Another one of the major problems with eating between meals is the kind of foods we eat between meals and the way we eat them. Whereas there is time during a proper meal for healthy, balanced foods, a snack is almost by definition something that's pre-prepared and unhealthy. Children should already be eating a built-in snack after school or in the afternoon; just make sure it consists of high-taste foods like fruits, yogurt or a French favorite like bread with a small piece of dark chocolate.

Once you're at the dinner table with foods you've chosen, don't limit children's access to those foods. You may want to stipulate that they eat a little more of everything rather than gorging on more chicken or pasta, but remember that if you're going to teach them that food is eaten at the table with the family and not in the afternoon in front of the refrigerator, they need to see that they can eat their fill in these settings. Naturally, the same doesn't necessarily go for dessert, but one way around this is to serve only small group portions of dessert. Opt either for a dessert that's small overall, like a small pie that feeds only the number of people you have at the table, or serve only a portion of the overall dessert, leaving the rest for the next day. Don't even bring the entire dish to the table.

The French philosophy about meals is that the main meal is eaten at lunchtime, and dinner is light. This has this been proven to be better for digestion, sleep and overall health. If you can do this, at least on weekends, it's worth trying. The French also space out their meals a lot more than we Americans do, eating late lunches and dinners. The earliest dinner for adults eating without children in France is at eight PM, meaning that children usually eat around seven if the parents are going out for dinner or having guests, or the whole family eats around seven or so together. This habit of eating later often means that the whole family can be together on a regular basis for dinner. Of course, depending on your particular work schedule and your children's bedtimes, seven o'clock may be way too late, but why not start out by trying to eat half an hour later instead?

All the Ingredients for Success

"Who wants to try his first oyster?" my sister-in-law sang out as her children filed into the dining room. That afternoon, she'd driven her children the 30 minutes to the Normandy seashore in order to buy the oysters, fresh for dinner. Her children had marveled at the different sizes and types, examining their hard shells and weighing them in their hands.

"I do!" shouted her seven-year-old, the eldest.

"Me too," chimed in her second, a five-year-old.

The other children looked at each other and shook their heads. No, not this time. With great fanfare, she served one oyster to each of her first two children, a look of pride spreading across their faces at being able to taste such a "grown-up" dish.

"Here, take some lemon. Oysters taste best with a little lemon. And, here's some bread." She handed them each a piece of the bread they had made together as a family the day before. "Should I put the butter on for you?"

With the careful look children have when they know they are doing something very important, Gaëtan and Hugues prepared their oysters and slurped them down.

"So, what did it taste like?"

"Like the sea! It's salty and slippery."

"Yum. Can I have another one?"

Teaching your children about taste should be a fun, family affair. Whether you're eating high-taste foods that explode with flavor like radishes or yellow peppers, more complex tastes like Dover sole with lemon or even good old family favorites like mashed potatoes and pork chops, take a moment to revel in the experience with your children. Show them that eating is enjoyable and interesting.

Everywhere in the world these days, children are exposed to bad food influences and it is our role as parents to provide them with an internal ability to appreciate food that will be useful for them their whole lives, much like you try to provide them with a moral compass that will guide them in difficult situations when you are not there to help.

Learning to taste and savor food can be fun, both for you and your children. (You may even find you develop better food habits yourself!) The French method, particularly now that it has been formalized, shows that keeping your child thin and healthy is really just a matter of changing a few behaviors and devoting a little more time and energy to meals. The key is to keep in mind that eating is enjoyable, particularly as you experience new, good-tasting foods as a family. Pass this unstressed attitude on to your children by helping them develop their own appreciation of food.

One Last Word

It's great to grow up American. Nowhere else in the world are children so exposed to such powerful ideas as independence, self-reliance and entrepreneurship along with that pick-yourself-up-by-your-bootstraps way of pushing ahead no matter what the circumstances. Living in France where over two thirds of college seniors say they want to spend the rest of their lives working 35 hours a week as government functionaries, I appreciate all the more the American character and will make sure that my children inherit it. Nonetheless, the French, undeniably, have a few things to teach us about good living, including the ability to appreciate food and taste.

With all the books focused on increasing your child's intelligence and chances for success (and no doubt his or her stress level), I wanted to write one about increasing your child's pleasure in living. Learning to taste both helps children to avoid the traps that Americans often fall into which create anxiety about food while providing them with the skills to appreciate and savor food, an activity most of them will engage in about three times a day for the rest of their lives. If you do choose to teach your child to taste, I have just one more thing to say: good luck, have fun and eat up!

More fun and games

Don't stop now! Once you've set the stage for your children to start appreciating their food, take it one step further by running your own taste classes at home. Choose a day, perhaps Sunday afternoon or whenever you have an extra hour or so, and get started. Another idea is to organize an informal taste class among several friends and their children. Take turns running the class and adapt the lessons to your children's own needs.

Here are a few more taste exercises to try with your children. When you've done them all, develop your own! Remember a taste class can be as simple as introducing your child to a new food, cuisine or way of eating something that's already familiar.

<u>The Tongue</u>
Ages: 3-5
Main idea: Teach children how taste works.

Preparation:

- A magnifying glass
- A mirror

Let the children examine each other's tongues to see each other's taste buds (or their own using a mirror). Discuss how they are the same or different. Ask them to draw what they see.

Supertasters
Ages: 7-12
Main idea: See if you or your child is a supertaster.

Preparation:

- A reinforcement ring from a three hole binder
- Blue food coloring
- A magnifying glass
- A mirror

Put the ring near the front of your child's tongue and die the part of the tongue inside the ring blue. Count the pink spots you see (each of which contains about six taste buds on average, but you can't see them).
More than 40 taste buds = supertaster
Between 20 and 40 = normal
Fewer than 20 = someone for whom most things taste bland, or who does not have favorite foods, otherwise known as a "non-taster"

More Fun and Games

<u>The Exotic Candies Game</u>
Ages: 7-12
Main idea: Show children that visual information is not always reliable.

Preparation

- If you have a Japanese confectionary anywhere near you, or you can get your hands on Japanese candies somewhere else, they are ideal for this exercise. Japanese candies are beautifully crafted, delicious looking bon bons, often filled with bean paste, a food that disappoints most American palates. Certain Mexican candies, like lemon-flavored salt, which comes in brightly colored plastic container that is naturally attractive to children, will also be good for this exercise. If you can't find either, whip up a bad-tasting cookie – well decorated with candies on top, but made without sugar.
- Home made fudge

Lay out your Japanese candies next to your homemade fudge (being careful to make the fudge look as unattractive as possible.) Ask your children which one they prefer and allow them to taste both. Show them that judging foods only on visual information (the primary method children use to decide whether they will eat something or not) is not always going to lead them to the best taste.

Apples
Ages 3-10

Main idea: Show children the range of different tastes available from one basic food, the apple.

Preparation

- You'll need as many different types of apples as you can find: Granny Smith, Red Delicious, Rome Beauty, Cortland, Empire, or any other of your favorite varieties. (Fact: 2,500 varieties of apples are grown in the United States and 7,500 around the world. Try to do this exercise in apple season – if you can combine it with a trip to an orchard to pick your own, all the better!)
- Prepare apple sauce, apple juice (made from the fresh apples), baked apples

Cut the apples into slices and ask your children to taste each one and describe the major sensation they have. Let them pick a favorite type of apple. Then let them taste how the flavor changes with cooking or mashing.

Planning a meal
Ages: 5-10

Main idea: Teach children the basic structure of balanced meals

Preparation:

- Pad and paper to make a shopping list

Ask children to help you plan dinner. Give them a choice of meat, vegetable and carbohydrate and get them to be as specific as possible about how each should be prepared. Be ready to suggest several simple recipes or sauces. Go shopping with them to the supermarket, or ideally, a farmer's market or a local market and help them choose each ingredient and then prepare the meal together.

More Fun and Games

<u>The Sound of Food</u>
Ages: 5-10
Main idea: Learn to identify the sounds made by various foods

Preparation:

- Blindfolds or something to cover your children's eyes so they can't peak
- Foods that make noise, like a stalk of celery, a carrot, popcorn, crackers, a piece of crusty bread, a crisp apple, some fresh lettuce, a pickle, a fresh pepper – or any of your favorites. (Don't let your child see what you are preparing beforehand.)

Blindfold your child and let him or her guess what food is being eaten by its noise. Can he tell the difference between the sounds of the various foods? With older children, switch and let them eat some of the foods while you try to guess what they are!

<u>Making Mayonnaise</u>
Ages: 5-10
Main idea: Show children that a basic condiment like mayonnaise can be made quickly and help them to understand the ingredients.

Preparation

- The yellow of one egg
- One teaspoon of mustard
- Salt and pepper
- Oil

Show children how to crack an egg to separate the yellow part and mix it with mustard, salt, pepper and a little bit of oil in a bowl. Let them use a whisk to whip the mixture into a sauce. Gradually add more oil until the sauce is thick.

<u>Farm store</u>
Ages: 5-12
Main idea: Teach children how naturally grown foods in season can taste better than their supermarket equivalents.

Preparation

- Find a local farm store near you

Depending on the season and the selection at the farm store, choose a variety of products. Some good foods if they are available are locally-made butter or cheese, fresh fruits such as peaches or apples, exotic mushrooms, homemade apple cider or lemonade and freshly-made donuts. Next, go to your local supermarket and buy your comparison foods: Factory-made butter, fruits perhaps bought from far away which have traveled for days or weeks, pre-packaged donuts, etc.

Ask your child to taste each of the farm store foods together with the store-bought ones and talk about the differences:

- How do the colors compare?
- Which smells fresher?
- Are there differences in texture?
- Is one juicier, creamier, moister, more firm than the other?
- Which one does your child prefer?
- What was more fun: going to the farm store or the supermarket?
- What is the season for (the fruit or vegetables you've chosen)?
- How is the (butter or doughnuts) made? Were they able to see some of the foods being made?

More Fun and Games

<u>Making Whipped Cream</u>
Ages: 5-10
Main idea: Teach children that cooking can be easy and fun. Help them to talk about food and understand where it comes from.

Preparation:

- Lay out a bowl for each child (and for yourself) of strawberries (or your favorite fruit)
- One bowl of *crème fraîche*
- One large bowl of ice cubes
- A whisk
- Sugar

Start by asking your children if they know how whipped cream is made and let them guess by looking at the ingredients that you have laid out. Explain to them that using the cream and the whisk you add little bubbles of air to the mixture. The bubbles get wrapped up in a layer of cream and water which creates the texture of whipped cream that they know from a can or the shaving cream that Daddy uses in the morning to shave.

Take the bowl of cream, add a bit of sugar and put the bowl in the larger bowl of ice cubes. Start whipping the cream and let each child take a turn. Put on the strawberries and serve.

Next time you make whipped cream for dessert, let your children make it themselves!

Make Chocolate Mousse

Ages: 7-12
Main idea: Make a great dessert together.

Preparation:

- You'll need seven ounces (200 grams) of very dark chocolate
- Six eggs
- A pinch of salt
- And 20 minutes, plus at least 6 hours of time before serving

Start by melting the chocolate. One way to do this is to put it in a bowl and place the bowl in a pot of simmering water. Stir occasionally until the chocolate becomes a thick liquid. Take out the chocolate in its bowl and let it cool or cool it down manually by running cold water on the outside of the bowl.

Break the eggs, putting the egg yellows in a large bowl and the egg whites in a mixing bowl. Add the pinch of salt to the whites and beat them into foam.

Pour the chocolate on the egg yellows and mix. Then use a big spatula to slowly incorporate the egg whites into this mixture, one spatula-full at a time. Delicately turn the chocolate mixture over the fluffy egg whites, being careful not to make them lose their foam.

Smooth out the top of the mousse, cover it and put it in the refrigerator for at least six hours. Take out of the refrigerator immediately before serving.

Variations: You can also scoop the mousse into some sturdy wine glasses to create individual mousses. Also try adding some fresh berries, like raspberries or strawberries to the top of the mousse.

Lightning Source UK Ltd.
Milton Keynes UK
UKHW010732250922
409416UK00001B/10